Safe or Not Safe

Paul Pechan · Ortwin Renn · Allan Watt
· Ingemar Pongratz

Safe or Not Safe

Deciding What Risks to Accept
in Our Environment and Food

 Springer

Paul Pechan
Institute of Communication and Media
Research
Ludwig Maximilians
University Munich
Oettingenstr 67
80538 Munich
Germany
paul.pechan@ifkw.lmu.de

Ortwin Renn
Dialogik
Lerchenstr 22
70176 Stuttgart
Germany
renn@dialogik-expert.de

Allan Watt
NERC Centre for Ecology and Hydrology
Bush Estate
Penicuik EH26 0QB
United Kingdom
adw@ceh.ac.uk

Ingemar Pongratz
Karolinska Institutet
Stockholm
Sweden
ingemar.pongratz@ki.se

ISBN 978-1-4419-7867-7   e-ISBN 978-1-4419-7868-4
DOI 10.1007/978-1-4419-7868-4
Springer New York Dordrecht Heidelberg London

Library of Congress Control Number: 2011926227

Printed on acid-free paper

Springer is part of Springer Science+Business Media (www.springer.com)

Introduction

Risk and uncertainty. These terms are very often mentioned in the media. But what is their real meaning? Are they important for our daily life? This book explains these terms in the context of our food chain. It provides a number of food-related examples of how scientists measure, assess, and manage potential risks and how this helps us to decide what is safe and not safe. The examples have one thing in common: They deal with problems that are man-made.

The first chapter sets the stage: Risk examples are given from our daily life, illustrating what we need to know about risks and how to manage them. This knowledge forms the basis to guide you through the three remaining thematic chapters, where more specific topics and examples are discussed.

The second chapter looks at biodiversity. Almost two million species have now been identified and the actual number of species in the world is estimated to be between 10–30 million. This enormous biodiversity is an essential provider of eco-system goods and services to our society. However, despite the important role biodiversity plays in our lives, all species that together comprise biodiversity face risk.

The third chapter investigates the risks chemical contaminants pose to our health through the food chain. The issue of food consumption has evolved from a relatively short chain of trading between producer and consumer to a complex chain where different parties are involved. Today, food consumption includes large-scale production, time-efficient handling, transport, and packaging of food. Along this chain, there are many possibilities to introduce unwanted chemicals into the foods we consume, sometimes with harmful outcomes.

The last chapter looks back at the nearly three decades of developing genetically modified foods. However, the technology used to create such foods has not been readily accepted by the public. What are the benefits and risks of this technology? The chapter not only looks at the science behind the technology but also addresses issues as consumer choice and rights, politics, product usefulness, and availability of alternatives.

This book is aimed at educators in formal and informal educational settings as well as interested general public.

About the Authors

Paul Pechan is a researcher at the Department of Communication and Media Research, Ludwig Maximilians University Munich. His background is in molecular biology, but since 2000 he has been involved in the research and communication of science to the public and youth. His latest work includes scientific papers on information recall, texts for 5 health films and a UNESCO manual on GMO for teachers.

Allan Watt is a research scientist at the Centre for Ecology and Hydrology, Edinburgh. He has led several international projects with a focus on conflicts between human activities and biodiversity.

Ingemar Pongratz is group leader and holds an Assistant Professorship at the Department for Biosciences and Nutrition at KI. Dr Pongratz's main research area is the biology and function of pHLH-PAS proteins, and he leads his own research group studying the effects of dioxin exposure at the cellular level.

Ortwin Renn serves as full professor and *Chair of Environmental Sociology and Technology Assessment* at Stuttgart University (Germany). He directs the *Interdisciplinary Research Unit for Risk Governance and Sustainable Technology Development (ZIRN)* at Stuttgart University and the *nonprofit company DIALOGIK*, a research institute for the investigation of communication and participation processes in environmental policy making.

Contents

Chapter 1
Risks

**Ortwin Renn, Julia Ortleb, Ludger Benighaus,
and Christina Benighaus**

1 Why Are We Writing About This Topic?

Risk is our daily work and often our obsession – as risk researchers we are working on research projects which are intended to increase our knowledge about all aspects of risks. Being a "risk researcher" means looking at things through a specific perspective – the perspective of what negative consequences a natural or man-made event, a technology, a decision could probably have on the world we live in. Our perspective is a socio-scientific one. This means, we are analyzing what consequences do risks have on the society and what can we do to decrease or prevent them. This includes the possible actions of a single consumer as well as strategies of whole governments to manage risks. How are risks perceived by people? What kinds of knowledge are needed to deal with different kinds of risks? Who should be involved, and when? What to do if conflicts evolve about how to handle risks? How to communicate risks? It is our job to answer questions like these. We are dealing with these questions in many different thematic areas: food safety, climate change, chemicals, nanotechnology, electromagnetic fields, etc. These risks pose very different problems and it seems difficult to find general strategies to deal with them.

This first part of the book is meant to set the stage for the following chapters: We want to convey insights into current risk research on a general level, before diving into the more thematically specialized chapters of the book. This means, we illustrate what you need to know on risks and how to handle them with examples of our daily life and give you a broad picture of the different aspects of risk research. This knowledge will form the basis to guide you through the three remaining thematic chapters, where more targeted strategies of dealing with different types of risks are presented.

The following section of the chapter will inform you about what risk is and what characteristics, concepts, and perceptions of risk exist. Section 3 introduces

O. Renn (✉)
Dialogik, Lerchenstr 22, 70176 Stuttgart, Germany
e-mail: renn@dialogik-expert.de

P. Pechan et al., *Safe or Not Safe: Deciding What Risks to Accept in Our Environment and Food*, DOI 10.1007/978-1-4419-7868-4_1,
© Springer Science+Business Media, LLC 2011

an integrated concept of how to deal with risks to which modern societies are exposed and explains, the concept of "risk governance." Section 4 deals with problems arising though conflicting views, values, and knowledge gaps in the risk field. The last section identifies the condition for handling and managing risks more effectively, efficiently, and in accordance with democratic principles.

2 Risk as a Science Topic and Expected Impact on the Society

2.1 What Is Risk?

Health risks are front-page news. Be it BSE, surface ozone, or radiation from transmitter stations of mobile phones, the popular press puts out a constant stream of risk warnings and sensational reports. The recent risk-related food scandals from BSE to Acrylamide provide ample evidence that there is no simple recipe for understanding and managing risks. When we talk about risks, we may associate many different things: fears of specific hazards such as a terrorist attack, concerns regarding potential failures of complex technological systems like the ones we might face with nuclear energy systems, uncertain projections regarding financial gains or losses that we may experience in the stock market, worries about natural disasters such as the tsunami in South Asia in 2004, but also the thrill of adventure produced through bungee jumping or other extreme sports. Included in the portfolio of risk may also be worries about the competence and trustworthiness of those who manage these different types of risks (Jaeger et al. 2001: 16f.).

In view of worldwide divergent preferences, variations in interests and values and very few if any universally applicable moral principles, risks must be considered as heterogeneous phenomena that preclude standardized evaluation and handling. At the same time, however, risk management and policy would be overstrained if each risky activity would require its own strategy of risk evaluation and management. What risk managers need is a concept for evaluation and management that on the one hand ensures integration of social diversity and multidisciplinary approaches, and, on the other hand, allows for institutional routines and standardized practices. This chapter provides a concept of how to understand, assess, and manage risks with special reference to food safety and biodiversity.

The concept of risk can thus be understood as a kind of perspective to analyze the uncertain consequences of future developments and changes in societies. Risks are like a pair of "glasses" through which the modern world is looked at. As the world has experienced a fast rush of major changes in the last decades, an abundance of risk-related scandals and debates show that risk has become a predominant topic in modern societies (Beck 1986).

Three actual examples have been chosen to be analyzed in depth as case studies in the following chapters of this book:

1. The *loss of biodiversity* as a consequence of the global demographic and technological development

2. Food safety issues on the example of Dioxin TCDD in Baltic sea fish
3. The potential risks of *genetically modified food* to human health

To give an impression on the variety and diversity of risk issues that have to be handled, here are some additional examples of large-scale disasters that have dominated the headline news over the last years:

- The terrorist attack of September 11, 2001
- Natural hazards like the devastating tsunami on Christmas 2004
- The hurricane Katrina in 2005
- The appearance of new infectious diseases like the severe acute respiratory syndrome (SARS) and avian influenza
- Food scandals like bovine spongiform encephalopathy (BSE)

Many definitions of the term *risk* exist and are used by various disciplines and for various risk events: One of the most common ones goes back to the 1980s, and in this definition the term risk denotes the possibility of adverse effects from some action or event with respect to something that humans value (Kates et al. 1985: 21; Fischhoff et al. 1984; see also Renn 1992). This definition combines two dimensions: the likelihood or chance of potential consequences on the one hand, and the severity of these consequences, due to human activities, natural events or a combination of both, on the other hand. This definition implies that the concept of risk does not exclusively describe negative consequences. The judgment whether the implications are positive or negative depends on the values that people associate with them. If we think, for example, of climate change as a risk, the possible consequences like global warming might be perceived differently by different people. Northern Europeans might have a more positive view as they would profit from minor temperature increases as they could increase agricultural productivity and tourism, while people from Africa or Asia are already suffering from lower agricultural productivity and an increase in natural disasters like droughts, floodings, etc.

Risk needs to be distinguished from the term "hazard," for which no common accepted definition does exist as well. *Hazards* describe the potential for harm or other consequences of interest (IRGC 2005: 19). A hazard can hence be the potential of a specific dose of a chemical to produce harm. The difference between risk and hazard is that as long as nobody is exposed to the chemical or an agent like, e.g., acrylamide, there is no risk, only the potential for harm. Probability and exposure are characteristics of the risk. Renn provides a useful conceptual distinction of the two terms: "hazards characterize the inherent properties of the risk agent and related processes, whereas risks describe the potential effects that these hazards are likely to cause on specific targets such as buildings, ecosystems, or human organisms and their related probabilities" (ibid.).

In both natural science and engineering, *risk* is further qualified as the mathematical product of likelihood of occurrence and severity of impact, resulting in a mathematical probability function applied across the range of potential damages. Why is this mathematical definition of risk used in most scientific disciplines? Science is based on the principle of intersubjective validation. This means, it must be both scientifically validated (i.e., other scientists must be able to verify the results when using the same methods) and expressed in numerical terms. Apart from the traditional elements of likelihood of occurrence and severity of damage with respect to health risks, risk taking also involves other risk-related and situation-related circumstances: For example, some components of risk are not covered in the traditional technical definition of risk[1]:

– The uncertainty that remains after assessing probabilities and potential for harm (there is, for example, still uncertainty about the long-term health effects of electromagnetic fields from mobile phones, etc.).
– The ubiquity describes the geographical spread of a damage.
– Persistence means the time, how long a damage lasts. The persistence of harmful effects is independent of their severity, even effects that seem not to be severe at first sight can turn out to be problematic due to their spread in time. For example, some chemicals, which do not seem to have severe negative effects at first sight, can turn out to be accumulating in the organism over years due to their persistence.
– Delayed effects over time, meaning that some risk consequences do not emerge immediately, but they appear after months or even years. This has, for example, been the case with the health effects of asbestos.
– The scope for institutional risk management and limitation (the range of possible management options can be limited by financial, political, or cultural reasons).

The technical concept of risks should not be confused with how individuals and social groups define and perceive risk. Many risk-related and situation-related factors play an important role in how risk is perceived by individuals, groups, or social institutions, which form the subjective factors in risk perception (Slovic 1987; Rohrmann and Renn 2000). The way risk is perceived differs, for example, depending on whether or not the individual has self-perceived control over the degree of risk involved with respect to personal control and management potential (Sjöberg and Drottz-Sjöberg 1994). Such subjective factors should not be deemed irrational. When we assess risk, it really does make a difference whether one can personally control the degree of risk (say, during leisure activities) or whether one must passively accept a given risk (e.g., passive smoking).

Risks consequently have to be understood as permanent companions of everyday life. As long as people value certain things or conditions and as long as they take decisions in the presence of uncertainty, they will face risks. Risks are hence a basic constituent of life.

2.2 Varying Concepts of Risk

When looking at risk, different disciplines and perspectives can be distinguished. These perspectives are listed below.

---

[1] The German Advisory Council on Global Change (WBGU) recommended a classification system based on seven generally applicable risk factors to define various types of risk. For each type of risk, a separate strategy was developed for assessment and management of those risks (WBGU 1998).

- *Technical concept*: this concept of risk, which is predominantly used by insurances, is expressed through the mathematical function of probability and harm. Harm refers to human health, environment, and capital assets.
- *Economic concept*: expresses risks in expected utilities, which can be losses or gains, and allows therefore a comparison between risks and benefits by weighting possible costs by the probability of their occurrence.
- *Ecological concept*: understands risks as a threat to ecosystem stability and sustainability.
- *Psychological concept*: subjectively expected utilities (based on individual perceptions of harm and likelihood and other qualitative factors such as contextual variables) are used by individuals to deal with risks.[2]
- *Sociological concept*: is a patchwork of different concepts, which have in common that they deal with social constructions of pending threats to all aspects of what individuals and groups value.
- *Cultural concept*: this concept deals with culture-specific rules and procedures for framing, analyzing, managing, and handling threats to society. Certain values are the basis. It works with mind-sets of individuals that are structured by cultural patterns.

All these concepts of risk emphasize different aspects of the risk phenomenon. They focus either on the type of harm or the qualification of uncertainties and ambiguities. In particular, the concepts differ in their approach or measure of uncertainty, in their definition of what constitutes undesirable outcomes and in their understanding of reality (ibid. 58). As a consequence, the different phases of risk governance need not only have to address the challenges outlined above, but also the varying concepts of risk in the different scientific disciplines.

2.3 Basic Components of Risk

For the analysis of traditional or systemic risks it is helpful to decompose the knowledge base of what we call risk into three major components. These components are *complexity*, *uncertainty*, and *ambiguity* (Klinke and Renn 2006).

2.3.1 Complexity

Often it is difficult to identify and quantify causal links between a multitude of potential causal agents and their specific adverse effects. The nature of this difficulty may be traced back to a number of different factors, which are subsumed under the term complexity: interactive effects among the causal agents (mutual strengthening

---

[2]This aspect is further developed in Sect. 5 on risk perception.

or weakening), positive and negative feedback loops, long delay periods between cause and effect, interindividual variation and intervening variables. These are only some of the multiple factors which give hints at complexity. It is precisely these factors that make high-level scientific investigations necessary, since the cause–effect relationships in complex risks are neither obvious nor directly observable. The global decrease in biodiversity is an impressive example for a risk that is characterized by high complexity. There are many factors, like the destruction of natural habits of endangered species, increasing land use for housing and industry, landscape fragmentation, intrusion of invasive species caused by globalized transport and travels, climate change, and environmental pollution, of which the interdependencies cannot completely be identified or quantified.

2.3.2 Uncertainty

This term describes a state of knowledge in which the likelihood of any harmful effects or even these effects themselves, cannot be precisely described, although the factors influencing the issues are identified. Uncertainty is different from complexity, but is often a result from an incomplete or inadequate reduction of complexity in modeling cause–effect chains. It comprises different components such as statistical variation, measurement errors, ignorance and indeterminacy (van Asselt 2000). All of these have one feature in common: uncertainty reduces the strength of confidence in the estimated cause–effect chain. If complexity cannot be resolved by scientific methods, uncertainty increases. But even simple relationships may be associated with high uncertainty if either the knowledge base is missing or the effect is stochastic by its own nature.

Uncertainty can be further disaggregated into separate components. Two epistemic components are "target variability," meaning differences in the vulnerability of targets (e.g., the different reaction of male and female organisms on medication) and "systematic and random errors in modeling," which are mainly driven by extrapolation (e.g., from animals to humans or from large doses to small doses). In these cases, uncertainty can be reduced through the generation of new knowledge or the advancement of present modeling tools.

One example of uncertainty can be found in the food sector, especially in the food-supplier-chain, and the possible contamination through chemicals. It is estimated that around 70,000 chemicals do exist in the environment, and every consumer is exposed to them, for example, through the food chain. The exact effects of every single chemical are yet not well known and most foods contain more that one chemical at the time. This means that often cocktail effects can be observed, of which the consequences are often unknown.

Other components of uncertainty cannot be reduced because they are aleatory, i.e., driven by chance. These components are "genuine stochastic effects," "system boundaries," and "ignorance or nonknowledge" (IRGC 2005: 30). An actual example is the risk of an uncontrolled spreading of genetically modified plants in the environment.

2.3.3 Ambiguity

The existence of different (legitimate) interpretations based on identical observations or data assessments is called ambiguity. Most of the scientific disputes in risk analysis do not refer to differences in methodology, measurements or cause–effect functions, but to the question of what all this means for human health and environmental protection. An example: Emission data of greenhouse gases is hardly disputed. Most experts debate, however, whether a certain emission constitutes a serious threat to the environment or to human health. Ambiguity may come from differences in interpreting factual statements about the world or from differences in applying normative rules to evaluate a state of the world. In both cases, it exists on the ground of differences in criteria or norms to interpret or judge a given situation. High complexity and uncertainty favor the emergence of ambiguity. On the other hand, there are also quite a few simple and almost certain risks that can cause controversy and hence ambiguity. This is, for example, the case in the discussion of speed limits on German motorways in order to reduce the risk of accidents.

Ambiguity comprises two dimensions. One is *interpretative ambiguity*, which describes different interpretations about the implications of a given hazard. The associated question to this dimension is: What does an assessment result mean? A typical example for interpretative ambiguity is the risk of electromagnetic fields (EMF). Studies have shown that laypersons judge the risks concerning EMF differently and generally higher than experts.

The other dimension is *normative ambiguity*, and raises the question about the tolerability of the hazard. It is based on the idea that there are varying legitimate concepts of what can be regarded as tolerable, "referring, e.g., to ethics, quality of life parameters, distribution of risks and benefits, etc." (IRGC 2005: 31). For example, genetically modified organisms (GMO) encounter a high level of opposition in the area of food, but are widely accepted in the area of medical applications, because they are associated with the hope for health benefits.

2.4 Characteristics of Risks in the Modern World: New Challenges to Risk Governance

A number of *driving forces* have been identified which are shaping our modern world and have a strong influence on the risks we face (OECD 2003: 10ff.):

1. The demographic development
2. Globalization
3. The rapid technological change
4. Changes within the socioeconomic structures and global environmental change

The *demographic development*, including the increase of the world population, the growing population density, and visible trends toward urbanization, accompanied by significant changes in the age structure of most industrial populations have led to more vulnerabilities and interactions among natural, technological, and habitual hazards. Demographic changes are also partially responsible for the strong interventions of human beings into the natural environment. Human activities, first of all the emission of greenhouse gases like CO_2, may cause global warming. As a consequence, they place growing stress on ecosystems and human settlements. In addition, the likelihood of extreme weather events increases with the rise of average world temperatures. Furthermore, these trends toward ubiquitous transformation of natural habitats for human purposes are linked to the effects of economic and cultural *globalization*: The exponential increase in international transport and trade, the emergence of worldwide production systems, the dependence on global competitiveness and the opportunities for universal information exchange testify to these changes and challenges. In terms of risks, these trends create a close web of interdependencies and coupled systems. Small disturbances have the potential to strongly increase through all the more or less tightly coupled systems. They might cause very high damages.

The development of globalization is closely linked to *technological change*. The technological development of the last decades has led to a reduction of individual risk, i.e., the probability to be negatively affected by a disaster or a health threat (for example, think of the eradication of many diseases in industrialized countries), but it has increased the vulnerability of many societies or groups in society: Among the characteristics of this technological development are the tight coupling of technologies with critical infrastructure, the speed of change and the pervasiveness of technological interventions into the life-world of human beings. All aspects that have been described as potential sources of catastrophic disasters (Perrow 1992; von Gleich 1999, 2003). Very typical examples for the restricted controllability of technological complexity are nuclear power plants, as have shown the catastrophe in Chernobyl. The youngest incidents in two German nuclear sites have not led to catastrophes but were impaired through delayed communication and unclear responsibilities.

In addition to the technological changes, *socioeconomic structures* have experienced basic transitions as well. In the last two decades efforts to deregulate the economy, privatize public services and reform regulatory systems have changed the government's role in relation to the private sector which had major effects on the procedures and institutional arrangements for risk assessment and risk management. Attitudes and policies are increasingly influenced by international bodies with conflicting interests and increasingly by the mass media.

These basic developments have induced a number of *consequences*:

- An increase of catastrophic potential and a decrease of individual risk, associated with an increased vulnerability of large groups of the world population with respect to technological, social, and natural risks.
- An increase in (cognitive) uncertainty due to the growing interconnections and the fast global changes.
- An increased uncertainty about a change in frequency and intensity of natural hazards due to global change.

- Strong links between physical, social, and economic risks due to the interconnections of these systems.
- An exponential increase in payments by insurances for compensating victims of natural catastrophes.
- The emergence of "new" social risks (terrorism, mobbing, stress, isolation, depression).
- An increased importance of symbolic connotation and attenuation of risks.

These recent trends and consequences of risks to society have led to the creation of a new risk concept – the concept of *emerging systemic risks*. These are risks "that affect the systems on which society depends – health, transport, environment, tele-communications, etc." (OECD 2003: 9). More specifically, systemic risks means the fact that risks to human health and the environment are embedded in a larger context of social, financial, and economic risks and opportunities. Systemic risks are at the crossroads between natural events (partially altered and amplified by human action), economic, social and technological developments and policy-driven actions both at the domestic and at the international level (OECD 2003; IRGC 2005; Renn and Klinke 2004). The most typical example for a systemic risk is global climate change. While it is a natural development that the climate system changes over time (think of the ice ages, for example), the actual developments are influenced by the large and still increasing amounts of human emissions of green-house gases. This leads to effects in the natural, economic, social, and technical systems, as they are all dependent on the climate and interdependent to each other.

Systemic risks lead to new challenges for risk management and risk governance, because the threat they pose to mankind is new and challenging. The interdependency of the natural and human systems, which enable the survival of close to seven billions of men, has never been as high as today. This is why these new threats are in the focus of actual risk research. New solutions to deal with risks must be found.

Among the most pressing challenges are:

- Finding more accurate and effective ways to characterize uncertainties in complex systems. Often, uncertainties cannot be completely resolved due to the interdependencies and complexities that characterize systemic risks. These uncertainties can be of a different nature, sometimes it is not possible to calculate the probability of a harmful event, sometimes it is even not possible to know all the factors that influence such an event. Hence, uncertainty can range from a simple lack of data to complete ignorance of the coherences. These different types must be characterized and decision rules have to be found how to deal with them.
- Developing methods and approaches to investigate and manage the synergistic effects between natural, technological, and behavioral hazards. This regards the organization and management of the knowledge of experts from many different disciplines, and at the interface of scientists and decision-makers responsible to implement the solutions. More collaboration and interdisciplinary is needed to be able to face risks that threat all relevant systems.

- Integrating the natural and social science concepts of risks to deal with both physical hazards and social risk perceptions. It is no longer sufficient to base decision only on the physical characteristics of hazards. Risk perceptions and values of the public have a high impact on the tolerability and acceptability of the risks and the solutions found to deal with them. Solutions to handle systemic risks might increasingly intervene in the everyday life, lifestyle, and freedom of people, so their concerns and perceptions have to be included when making decision.
- Expanding risk management efforts to include global and transboundary conse- quences of events and human actions. Decisions that are taken within one country will, in the context of systemic risks, have consequences for other countries as well (e.g., as regards to increase or decrease of greenhouse gas emissions). This means, that more people have to be included into the decision- making processes, i.e., more governments, more stakeholder groups, etc. More co-operation is needed, while the cultural differences between countries have to be respected.

In Chap. 3, we will present a framework that promises some solutions of how to deal with these challenges. But before this framework is explained in more detail, it is necessary to categorize the risks further that we are covering in this book.

2.5   The Integration of Perceptions and Social Concerns

Why do we need to include risk perceptions and concerns into the governance of modern risks? Risk consequences are judged differently by varying actor groups or individuals, depending on their "perception" of the risk. It does make no difference whether these consequences are intended or unintended. As the validation of the consequences depends on differing values and perceptions, risks can be described as mental or social "constructs" (OECD 2003: 67). This leads to:

- Different individual judgments about the severity and probability of risks
- Conflicts about how to handle them correctly
- The assessment if the measures are taken are acceptable, tolerable, or intolerable

"*Perceptions*" can be understood as the different images or mental models that are associated with risk by different cultures, groups, or individuals. It is these perceptions, i.e., what humans perceive of the world and what attitudes they develop toward it, that drives their behavior, not scientific facts. They result from common sense reasoning, personal experience, social communication, and cultural traditions (IRGC 2005: 31; Brehmer 1987; Drottz-Sjöberg 1991; Pidgeon et al. 1992; Pidgeon 1998). From an evolutionary perspective, humans have been using relatively consistent patterns of coping with dangerous situations. They can be reduced to four basic instinctive strategies, based on their perception of the risk: "*flight, fight, play dead* and, if appropriate, *experimentation* (on the basis of trial and error)" (IRGC 2005: 31).

As the nature of risk has changed with the growing complexity of the world (Sect. 2), these basic, instinct driven patterns of risk perception have been enriched by cultural and social influences. These perceptions influence the estimations and acceptability or risks and play therefore an important role in contemporary risk governance. Today, there exists a variety of scientific approaches that deal with risk perception, using different perspectives and concepts.

One of the initial concepts of "perceived risk" was first established by the psychologists Fischhoff, Lichtenstein, and Slovic in 1978.[3] This concept is known as the *"psychometric approach"* and uses qualitative evaluation patterns that go beyond the technical factors that are usually used by risk assessors, i.e., occurrence probability and extent of damage. Here, two classes of qualitative perception patterns are used: risk-related patterns (which refer to the properties of the source of the risk, e.g., the perceived "dread" of a consequence or if a risk is known or unknown to the observer) and situation-related patterns (which refer to the pecu-larities of the risky situation, e.g., voluntariness of exposure to a risk, controllability, or distribution of risks and benefits) (IRGC 2005: 32; Fischhoff et al. 1978; Slovic 1987, 1992). The psychometric approach is based on four intentions:

- To establish "risk" as a subjective concept, not an objective entity
- To gain a better understanding of the cognitive structure of risk judgments, usually employing multivariate statistical procedures such as factor analysis, multidi-mensional scaling or multiple regression
- To add social/psychological aspects to risk assessment and management
- To accept preferences of "the public" (i.e., lay people, not experts) as additional yardsticks for evaluating risks

Based on psychometric studies, a new concept of classifying risk perceptions has emerged which is referred to as *"semantic risk patterns."* Five patterns can be described (Renn 2004; IRGC 2005: 32):

Pattern 1: Risks posing an immediate threat (e.g., nuclear energy or large dams)
Pattern 2: Risks being understood as a blow of fate (e.g., natural disasters)
Pattern 3: Risks presenting a challenge to one's own strength (e.g., risky sports activities like freeclimbing)
Pattern 4: Risks as a gamble (e.g., lotteries, stock exchange, or insurances)
Pattern 5: Risks as an early indication of insidious danger (e.g., food additives, ion-izing radiation, viruses)

These semantic patterns help individuals to deal with new situations by associating them to similar and therefore "known" patterns. As an example, *genetically modified tomatoes* would be subsumed under the pattern "risk as an early indication of insidious danger." This risk could be described by high levels

---

[3]For a comprehensive review and documentation of this body of research see Rohrmann (1995), overviews are provided by Fischhoff et al. (1993), Guerin (1991), Jungermann and Slovic (1993), Pidgeon et al. (1992), and Renn (1986, 1990).

in the characteristics involuntariness, unknown risk, and a perceived low level of personal or institutional controllability. Together with risks of the first category (related to a very high level of "dread"), these types of risks are confronted with the danger of stigmatization[4] and lead therefore very often to low levels of tolerability and acceptance. Another example is the loss in biodiversity. This can also be placed in pattern 5. The risk has already taken worldwide dimensions but proceed continuously without a major event.

The described approaches show that the acceptability of a specific risk does not only depend on its level of occurrence probability and the extent of damage, but also on a number of qualitative characteristics that influence risk.

3   Analysis of the Risk Issues Involved

3.1   How to Deal with Systemic Risks?

We have learned in the first two chapters that risks are getting more complex, uncertain and ambiguous in today's world, due to the described trends of the demographic development, globalization, technological developments, and the changing socioeconomic structures and that therefore qualitative risk characteristics, such as individual perceptions, have to be take into account when handling these systemic risks. Dealing with these risks and with the way their consequences are interlinked, is captured with the term "risk governance." "*Governance*" has gained considerable popularity in such different research fields as international relations, comparative political science, policy studies, sociology of environment and technology and risk research. It describes the structures and processes of collective decision making, including governmental as well as nongovernmental actors (Nye and Donahue 2000). On the global level, governance describes a horizontally organized structure of functional self-regulation encompassing state and nonstate actors bringing about collectively binding decisions without superior authority (Rosenau 1992; Wolf 2002).

"*Risk governance*" involves the "translation" of the substance and core principles of governance to the context of risk and risk-related decision making (IRGC 2005: 22f.). In relation to the challenges of modern systemic risks, this means that there is a need for an integrated analytic framework that incorporates the views and perceptions of the various actor groups and includes the integration of scientific, economic, societal, and cultural aspects of the risks.

---

[4] The concept of "stigma" cannot be treated here since this would exceed the scope of this document. For more information see Kunreuther and Heal (2003).

3.2 The "Traditional" Understanding of Risk Governance

The scientific preoccupation with risk governance has its roots in the traditional understanding of *risk analysis*. Being strongly based on natural science concepts with a technical understanding of risk (risk as product of probability of occurrence and degree of harm), three components of risk governance are traditionally differentiated:

- Risk assessment
- Risk management
- Risk communication

Risk assessment describes the tasks of identifying and exploring the types, intensities and likelihood of the (normally undesired, negative) consequences related to a risk. In most cases, the results are expressed in quantified terms. Consequently, risk assessment can be defined as a tool of gaining knowledge about risks and is mainly located in the scientific area. The aim of risk assessment can thus be identified to describe a risk as precisely as possible and, if appropriate, to quantify it (OECD 2003: 66). The main challenges during the risk assessment phase are high levels of complexity and scientific uncertainty.

For example, in the case of pesticide residues in food, the assessment of the health risk of the residues of a single pesticide is comparatively unproblematic – through the characterization of dose–response relationships. But the concomitance of the residues of multiple pesticides together with additional multiple stressors from the environment and the assessment of their combined effects on human health poses a problem to risk assessors because of the complexity of the dose–response relationships of multiple residues. Other examples are the uncertainty of the effects of genetically modified organisms (GMO) shown in the GM tomato case (book Chap. 4) or the complex interplay of factors that cause the decrease of biodiversity (book Chap. 2). As a consequence, the measurement, statistical description and modeling of such types of risks can pose serious problems to the risk assessors.

Risk management, on the other side, describes the task to prevent, reduce or alter the consequences identified by the risk assessment through choosing appropriate actions. Accordingly, it can be defined as a tool for handling risks by making use of the outcomes of the risk assessment process. This task is located in the area of decision-makers – mainly in the field of politics, but in the economic sector as well. The main challenge to risk management is the existence of ambiguity, as it concerns the interpretation of the scientific findings and judgments about the tolerability or acceptability of a specific risk. This is specifically true for the judgment of genetically modified foods and feeds.

The obvious distinction between risk assessment (the scientific knowledge related to a specific risk) and risk management (the decision making of how to handle risks) often becomes blurred, if one takes a closer look into the risk governance processes. While risk assessment concentrates on the risk agent or the source of the agent themselves, and tries to identify the extent of damage as well as the probability of its occurrence, risk management has to take into account a much wider field (IRGC 2005: 21; Stern and Fineberg 1996; Jasanoff 1986: 79f.; 2004). It comprises preventive as well as reactive action. But risk management depends on the knowledge input from risk assessment. This is a crucial point, because the outcome of the risk assessment phase might on the one hand be very directive, which means, leaving only one option for the action to be taken. This could, for example, be the case if the assessment of the health effects of a specific pesticide results in the finding that it is genotoxic already in very low doses, and the only option for preventing harmful health consequences is a complete ban of the product. If this is the case, decision making is already included in the risk assessment phase. On the other hand, risk management does not only have to consider risk assessment outcomes, but also might, for example, have to alter human wants and needs, e.g., to prevent the creation or continuing of the risk agent, or to suggest alternatives or substitutes to a specific risk agent. It can also comprise activities to prevent exposure to a risk agent by isolating or relocating it or take measures to increase the resilience of risk targets.[5] This means, the issues that have to be taken into account by risk managers are often going far beyond the direct consequences of a risk. The case of the regulation of genetically modified organisms illustrates this complex task: The risk managers do not only have to consider the possible negative health effects that might be a consequence of, e.g., the consumption of genetically modified food, but also indirect consequences like possible losses in biodiversity due to the spread of genetically modified species, ethical concerns raised by religious or moral beliefs regarding the principle of a fundamental manipulation of living organisms, or effects of the ban (or public funding on the other hand) on the competitiveness of the national economy.

Risk communication is the third key element in the traditional understanding of risk governance. Its task was initially defined as bridging the tension between expert judgment and the public perceptions of risks, which often vary to a large extend (Sect. 5).

The "Committee on Risk Perception and Communications" defines it, "as an interactive process of exchange of information and opinion among individuals, groups, and institutions. It involves multiple messages about the nature of risk and other messages, not strictly about risk, that express concerns, opinions, or reactions to risk messages or to legal and institutional arrangements for risk management" (US National Research Council 1989).

The communication studies distinguish models which analyze the communication processes between different institutions. As one of the first, Harold Lasswell (1948)

---

[5]Resilience in this context means a protective strategy to strengthen the whole system against consequences of a certain risk, to decrease its vulnerability. This strategy is mostly taken in the case of unknown or highly uncertain risks. A well-known example from the health system is the vaccination in order to strengthen the immune system. Other possible measures are to design systems with flexible response options, or to improve the emergency management (IRGC 2005: 79).

described the single elements of the communication process with one simple question: "Who says what in which channel to whom with what effect?" (Fig. 1.1).

This simple question was revived from Shannon and Weaver (1949) and transferred into a mathematical model. The linear model was actually designed for the fast transmission of electronic signals for the Bell Telephone Company. Because of the simple usage and the description of the communication process between encoder and decoder, the model was transferred into general communications studies and the analysis of risk communication, too.

The model from Shannon–Weaver (Fig. 1.2) is too static and shows only the linear or one-way-communication process. This can lead to false interpretations, because human communication cannot be defined as linear, but as action, reaction, acceptance, and attitude.

Schramm (1954) adds the feedback component to the traditional one-way-communication-model. This was the foundation of the two-way-communication-model (Fig. 1.3).

Following the model of Schramm (1954), a two-way-communication should be used, in which the risk communicator directly contacts the target group and collects

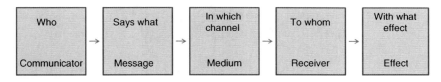

Fig. 1.1 Elements of the communication process (adapted and modified from Lasswell 1948)

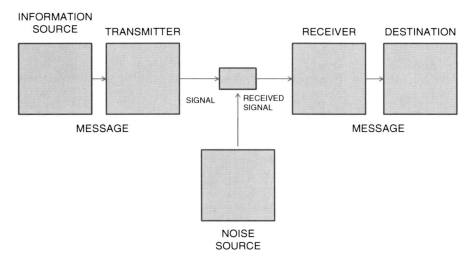

Fig. 1.2 Shannon–Weaver Mathematical Model (adapted and modified from Shannon and Weaver 1949)

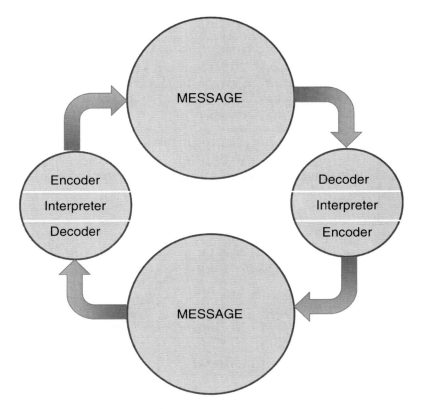

Fig. 1.3 Conservation model from Schramm (1954, adapted and modified)

their feedback. This target group should bring in arguments, ideas, impressions, judgments or statements (Renn and Kastenholz 2000: 30). Accordingly, after Schramm, the main characteristic of the two-way communication is the permanent transfer of the roles from being the sender or the active listener. Communication channels could be public events, forums, panels, exhibitions, printed material, or the internet, in which a feedback to the publisher is planned:

> Two-way communication is clearly a prerequisite for all forms of successful communication, but it is often hard to implement and requires flexibility and the willingness to adapt to public concerns on the side of the communicating institution (Renn and Kastenholz 2000: 30).

Actors on risk issues could access to strategies according to risk type and purpose. The essential element is an exhaustive analysis of the risk, and – similar to any other management process – the detailed definition of the goals and tasks.

In a review of risk communication approaches, William Leiss identified three phases in the evolution of risk communication practices (1996: 85ff.), which are briefly presented in Table 1.1.

Four major functions of risk communication and their goals can be identified (Morgan et al. 1992; OECD 2002; IRGC 2005: 55ff.). These four functions aim at

Table 1.1 Phases in risk communication

No.	Type	Characteristics
Phase 1	*One-way communication*	– Convey probabilistic thinking to the public
		– Application of risk comparisons
		– Educate the laypersons to acknowledge and accept risk management practices
		– Failed to convince audiences
Phase 2	*One-way-communication* Convey a persuasive message to the public	– Emphasize persuasion and focus on efforts of public relations to convince people that parts of their behavior were unacceptable
		– Some successes to change unhealthy behavior, but most people did not believe the messages
		– Altogether, this phase has had little effect
Phase 3	*Two-way communication* All members including the risk managers are involved	– To build up mutual trust by responding to the concerns of the public and relevant stakeholders
		– To assist stakeholders in understanding the rationale of risk assessment results and risk management decisions
		– To help stakeholders to make informed choices about matters of concern to them

Table 1.2 Functions and goals of risk communication

No.	Function	Goal
1	*Education and enlightenment*	Informing the public about risks, including risk assessment results and the handling of the risks according to risk management strategies
2	*Risk training and inducement of behavioral changes*	Helping people to cope with risks
3	*Promotion of confidence in institutions responsible for the assessment and management of risks*	Giving people the assurance, that those responsible for risk assessment and risk management act in an effective, efficient, fair, and acceptable manner
4	*Involvement in risk-related decisions and conflict resolution*	Giving stakeholders and representatives of the public the opportunity to participate in the risk-related decisions

helping all affected actors, i.e., stakeholders as well as the general public, to make informed choices when facing a risk (Table 1.2):

These four major functions pose a number of challenges to those responsible for risk communication (IRGC 2005: 57). They have to explain the concept of probability and stochastic effects to a broad audience. Otherwise, wrong interpretations of probabilities or exposure effects might lead to overreactions up to the stigmatization of a risk source (or to the opposite as well, as can be illustrated by the comparison of the risks of driving a car, which is often underestimated, and to travel by plane, which is most of the times overestimated). Dealing with stigmatized risk agents or with highly dreadful consequences is another challenge for risk communication. Risks, such as nuclear energy, can produce high levels of mobilization and very emotional reactions in the public. The example of the stigmatization of genetically modified food illustrates, that risk communication also has to take into account much more general convictions as well, such as ethical, religious, and cultural beliefs.

Risk communication, in this traditional understanding, is seen as a separate issue, which has as its main task to "educate the public" (IRGC 2005: 54), i.e., to communicate the results of experts' assessments to the wider public. In this understanding, risk communication follows the two phases of risk assessment and risk management, and is more one-way information than two-way communication, that takes into account varying perceptions and concerns.

The situation we are currently facing is a situation of change. The traditional risk analysis approach with its three described components is being increasingly criticized. In the view of a growing number of risk governance experts, in this "traditional" triangle, the *interfaces of the three components* risk assessment, risk management and risk communication are not adequately designed. The crucial point in the relationship of risk assessment and risk management is the general question of the influence of policy on science and vice versa. In the last two decades, the question was raised repeatedly of how to protect scientific risk assessment from inappropriate policy influences.[6] The institutional separation of these two tasks, like it has been implemented, for example, in the food sector, is a first step into this direction, but the implementation is still in a very early phase.[7]

This is why in the last years, a number of new models and approaches of risk governance have emerged resp. are emerging. These models are predominantly of theoretical and analytical nature. So the actual situation can be described as a situation of paradigm shift and the new models are currently in a phase of testing, improvement, and revision. One of these innovative models of risk governance is described in the following section.

3.3 *The Need for an Integrated Framework of Risk Governance*

The new challenges of systemic risks and recent tendencies in the handling of these risks, which have led to highly controversial conflicts about how to handle these risks, have shown that the three "generic" categories of risk governance, as they have been described above, are not sufficient to analyze and improve the risk governance processes. The characteristics of modern systemic risks (Sect. 4) require

---

[6] For the area of food safety, cf. Trichopoulou et al. (2000).

[7] For the area of food safety, cf. Dreyer et al. (2009).

new concepts, which are able to deal with the described challenges. This means, that besides the "factual" dimension of risk (which can be measured by risk assessors) the "socio-cultural" context has to be included as well, as systemic risks are characterized by affecting the whole "system" that humans live in.

The International Risk Governance Council (IRGC) has developed a proposal for an integrated framework for risk governance to help analyzing how society could better address and respond to such risks. To this end, the IRGC's framework maps out a structured approach which guides its user through the process of investigating global risk issues and designing appropriate governance strategies. This approach combines scientific evidence with economic considerations as well as social concerns and societal values and, thus, ensures that any risk-related decision draws on the broadest possible view of risk. The approach also states the case for an effective engagement of all relevant stakeholders.

Drawing on learning from a selection of current approaches to what has often summarily been termed "risk analysis" or "risk management," the framework offers a full risk handling chain ranging from how risk is identified, assessed, managed, and monitored to how it is communicated. This chain, which is in reality rarely sequential, breaks down into four main phases. The principal distinction between the knowledge gaining tool (assessment sphere) and the decision-making tool (management sphere) can still be identified. But there are also new elements, which combine these two generic steps.

The different components, which form the risk governance cycle, are briefly presented (Fig. 1.4).

The first phase, "*pre-assessment*" captures, and brings to the open, both the variety of issues that stakeholders and society may associate with a certain risk as well as existing indicators, routines, and conventions that may prematurely narrow down, or act as a filter for, what is going to be addressed as risk. It includes four elements: *Problem framing* describes the different perspectives on the conceptualization of the issue: the question of what the major actors (e.g., governments, companies, the scientific community, and the general public) select as risks. For example, is the global warming through climate change a risk, an opportunity or just fate? This element defines the scope of all the subsequent elements. *Early warning* comprises the institutional arrangements for the systematic search for new hazards. New phenomena such as, for example, the increase in extreme weather situations are taken as indicators for the emergence of new risks. *Screening* (or monitoring) describes the action of allocating the collected information on new risks into different assessment and management routes. This means, criteria like hazard potential, ubiquity, persistence, etc., are collected, systematically analyzed and amalgamated (Is the risk new? Is it an emergency? etc.) and related to potential social concerns. Finally, scientific conventions for risk assessment and concern assessment are defined (What methods will be used to assess the risk? etc.).

The second phase, "*risk appraisal*," provides the knowledge base for the societal decision on whether or not a risk should be taken and, if so, how the risk can possibly be reduced or contained. Risk appraisal thus comprises a scientific assessment of

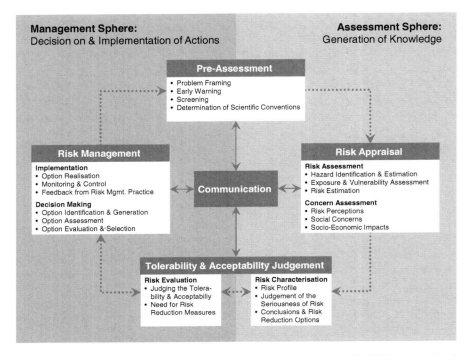

Fig. 1.4 IRGC Risk Governance Framework – General Model (IRGC 2005, adapted and modified, p. 365)

both the risk and of questions that stakeholders may have concerning its social and economic implications. This element consists of three generic components: *Hazard identification and estimation*, which describes the methods of recognizing the potential for adverse effects and for assessing the strength of cause–effect relationships. *Exposure/vulnerability assessment* defines the modeling of the diffusion plus the exposure pathways and the effects on the risk targets. In this step, those people are identified, that are (especially) affected by the risk, for example, people with a compromised immune system, very old and very young people, are vulnerable related to an influenza pandemia. The component *risk estimation* can be divided into two parts: quantitative estimation describes the probability distribution of adverse effects, while qualitative estimation comprises the construction of whole scenarios of combinations of different hazards, exposures, and qualitative factors. *Concern assessment* is also understood as a source of knowledge and includes the varying risk *perceptions and concerns* of all affected actors in the risk context (Sect. 5 on risk perception). *Socioeconomic impacts* and possible *economic benefits* are also considered in this step.

The third (and most controversial) phase, "*risk characterization and evaluation*" makes a judgment call on whether or not a risk is acceptable or – in view of the benefits it provides and if subject to appropriate risk reduction measures – at least tolerable.

Input for this decision comes both from compiling scientific evidence gained in the appraisal phase (risk characterization) and from assessing broader value-based issues and choices that also bear on the judgment (risk evaluation). Risk characterization includes the creation of a *risk profile* (including the outcomes of risk assessment), the *judgment on the seriousness of the risk* (including questions like: Are there effects on the equity of risk and benefits? Does the public acceptance exist?) and *conclusions and risk reduction options* (including suggestions for tolerable and acceptable risk levels). In the Risk Evaluation step, societal values and norms are applied to the judgment on tolerability and acceptability. In this step, the need for risk reduction measures is determined (this includes the choice of a specific technology, the determination of the potential for substitution, risk–benefit comparisons, die identification of political priorities and compensation potential, conflict management strategies, and the assessment of the potential for social mobilization). In this step in between scientific and policy-making contexts, the options for risk management are generated.

One possibility to classify risks is the *"traffic light model,"* a figure that is often used for classifying different natural and man-made risk areas. It supports assessment and management processes. This figure locates tolerability and acceptability in a risk diagram, with probabilities on the *y*-axis and extent of consequences on the *x*-axis (Fig. 1.5). In this variant of the model, the red zone signifies intolerable risk, the yellow one indicates tolerable risk in need of further management actions (in accordance with the "as low as reasonably practicable" ALARP – principle) and the green zone shows acceptable or even negligible risk.

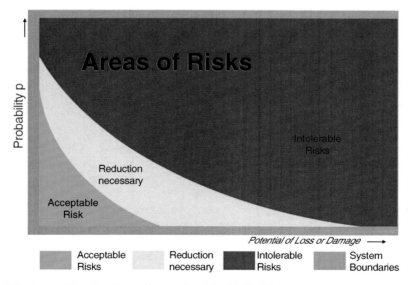

Fig. 1.5 Acceptable, tolerable, and intolerable risks (Traffic Light Model, adapted and modified, from IRGC 2005, p. 150)

This figure may help in situating risks within the dimensions of acceptability and tolerability by using e.g., psychometric characteristics or semantic patterns.

The fourth phase, *"risk management,"* designs and implements the actions and remedies required to tackle risks with an aim to avoid, reduce, transfer or retain them. Based on the development of a range of management options, risk management decisions are taken and put into practice. Depending on these outcomes, risk management has to fulfill two tasks: *Implementation* of the generated options includes the *option realization*, the *monitoring and control* of the consequences and the collection of *feedback from risk management practice*. The *decision making* includes option identification, generation and option assessment, and is accordingly interdependent with the tolerability and acceptability judgment step. The arrow between "Tolerability and Acceptability Judgment" and "risk management" goes into both directions. In most cases, the risk is only reduced, but will not reach the level Zero. After the analysis of the measures a second judgment might be necessary, in order to check if the risk is now acceptable.

The final element of the risk handling chain, *"risk communication,"* is of crucial importance in all phases of addressing and handling risk. It is placed in the center of the whole governance cycle. It should enable stakeholders and civil society to understand the risk itself and the rationale of the results and decisions from the risk appraisal and risk management phases when they are not formally part of the process. Even more importantly, when they are themselves involved in risk-related decision making, risk communication must also help them to make informed choices about risk, balancing factual knowledge about risk with personal interests, concerns, beliefs, and resources.

Risk communication has to deal with long-term and delay effects of risks, which often compete with short-term advantages in the view of different actor groups. Similar challenges are to provide an understanding of synergistic effects with other lifestyle factors or other risks and to address the problem of remaining uncertainties and ambiguities. The communication of such complex coherences demands a great deal of social competence, as it has to face the differing concerns, perceptions, and experiential knowledge of the different audiences addressed. On an international level, risk communication has additionally to cope not only with intercultural differences but with differences between various nations and cultures as well.

In this understanding, risk communication does not stand at the end of the risk governance process, but is an important element of all phases of the cycle. It is to be understood as a mutual learning process. The perceptions and concerns of the affected parties is meant to guide to risk assessors and risk managers in their selection of topics.

The framework is currently being tested for efficacy and practicability – i.e., can the framework help ensure that all relevant issues and questions are being addressed, and, does it support the development of appropriate risk governance strategies. Tests are conducted in the form of short case studies applying the framework to different risks, including those related to genetically modified organisms, stem cells, nature-based tourism and the European gas infrastructure. The results from these tests will serve as input to any necessary revisions to the framework.

3.4   Risk Management Styles According to Different Regulatory Styles of Risk Governance

Risk management has to cope with risks, which have been identified as simple or as problematic either due to complexity, to high unresolved uncertainty or due to ambiguity. Risk management strategies have to be adopted to these types of risk problems. The specific management strategy, the appropriate instruments and the degree of stakeholder involvement have to be chosen according to these criteria. But additionally, the governance process depends from the specific political culture that predominates in the corresponding region or, what is less obvious, in a specific risk domain (IRGC 2005: 61).

As far as the scientific input is concerned, a tendency to a development into the direction of an identical or at least similar language toward risk governance can be observed (Rohrmann and Renn 2000; Löfstedt and Vogel 2001). But additionally to scientific input, risk management is influenced by other components like systematic knowledge, legally prescribed procedures and social values (IRGC 2005: 62). This effects the outcomes of risk management. It may, for example, influence inclusion or selection rules, interpretative frames, or the handling of evidence.

Consequently, cultural diversity and the historical development of the political culture in the different countries have led to varying policy-making styles. They have, for example, influenced and shaped the relevant institutions. A number of common approaches for specific settings has been identified and is illustrated in Table 1.3 (IRGC 2005: 63).

Giving consideration to political and regulatory culture allows reference to how different countries or organizations within countries handle and regulate risks. Although management styles may become more homogeneous (particularly in industry), there is no common, global methodology in risk handling. The same risk may be processed differently and be subject to a different management decision depending on such factors as national culture, political tradition, and social norms. Accordingly, in some environments, a top-down ("vertical governance") approach will dominate; in others, an inclusive "horizontal governance" will be the norm.

4   Stakeholder and Public Involvement

4.1   Stakeholder Involvement and the Synthesis of Conflicting Perspectives

The risk governance process, as it has been described above, implies decision-making processes, which affect various groups of actors. On a general level, there is the distinction between the risk producers on the one hand, and those who are exposed to the risks on the other hand. It is obvious, that between these two groups, conflicting interests are to be expected. Both groups can be further divided into subgroups with

Table 1.3 Characteristics of Policy-making styles (*source*: IRGC 2005: 63)

Style	Characteristics	Risk management
1. *Adversarial approach*	• Open to professional and public scrutiny • Need for scientific justification of policy selection • Precise procedural rules • Oriented toward producing informed decisions by plural actors	• Main emphasis on mutual agreements on scientific evidence and pragmatic knowledge • Integration of adversarial positions through formal rules (due process) • Little emphasis on personal judgment and reflection on the side of the risk managers • Stakeholder involvement essential for reaching communication objectives
2. *Fiduciary approach (patronage)*	• Closed circle of "patrons" • No public control, but public input • Hardly any procedural rules • Oriented toward producing faith in the system	• Main emphasis on enlightenment and background knowledge through experts • Strong reliance on institutional in-house "expertise" • Emphasis on demonstrating trustworthiness • Communication focused on institutional performance and "good record"
3. *Consensual approach*	• Open to members of the "club" • Negotiations behind closed doors • Flexible procedural rules • Oriented toward producing solidarity with the club	• Reputation most important attribute • Strong reliance on key social actors (also nonscientific experts) • Emphasis on demonstrating social consensus • Communication focused on support by key actors
4. *Corporatist approach*	• Open to interest groups and experts • Limited public control, but high visibility • Strict procedural rules outside of negotiating table • Oriented toward sustaining trust to the decision-making body	• Main emphasis on expert judgment and demonstrating political prudence • Strong reliance on impartiality of risk information and evaluation • Integration by bargaining within scientifically determined limits • Communication focused on fair representation of major societal interests

distinct interests of their own, the so-called stakeholder. They are defined here "as socially organized groups that are or will be affected by the outcome of the event or the activity from which the risk originates and/or by the risk management options taken to counter the risk" (IRGC 2005: 49). In general risk issues affect the four main stakeholders in society. These are *political, business, scientific,* and *civil society* representatives (as far as they are socially organized). Additionally, other groups that play a role in the risk governance process, can be defined: the *media*, *cultural elites* and *opinion leaders*, and the *general public*, either in their role as nonorganized *affected* public, or as the nonorganized *observing* public (ibid.).

As governance aims at reaching acceptance of the outcomes of the decision-making process, the interests of all these different actors have to be met. At the same time, however, the number of options and the procedures how they are selected have to be restricted, as time and effort of the participants of the governance process have to be regarded as spare resources and therefore treated with care. Consequently, an inclusive risk governance process, as it is required when facing new risks, can be characterized by *inclusion* of all affected parties on one hand, and *closure* concerning the selection of possible options and the procedures that generate them, on the other hand.

Inclusion describes the question of what and whom to include into the governance process, not only into the decision making, but into the whole process from framing the problem, generating options, and evaluating them to coming to a joint conclusion. This goal presupposes that, at least, major attempts have been made to meet the following conditions (IRGC 2005: 49f.; Trustnet 1999; Webler 1999; Wynne 2002):

– Representatives of all four major actor groups have been involved (if appropriate)
– All actors have been empowered to participate actively and constructively in the discourse
– The framing of the risk problem (or the issue) has been co-designed in a dialog with the different groups
– A common understanding of the magnitude of the risk and the potential risk management options has been generated and a plurality of options that represent the different interests and values of all involved parties have been included
– Major efforts have been made to conduct a forum for decision making that provides equal and fair opportunities for all parties to voice their opinion and to express their preferences
– There exists a clear connection between the participatory bodies of decision making and the political implementation level

Two goals can be reached with the compliance of these requirements: the so-included actors have the chance to develop faith in their own competences and they start to trust each other and to have confidence in the process of risk management.

While these aims can be reached in most cases where risks are able to be governed on a local level, where the different parties are familiar with each other and with the risk issue in question, it is much more difficult to reach these objectives for risks that concern actors on a national or global level, and where the risk is characterized by high complexity or where the effects are, for example, not directly visible or not easily referred to the corresponding risk agent. Sometimes, one party may have an advantage from performing acts of sabotage to the process, because their interests profit from leaving the existing risk management strategies into place.

Consequently, inclusive governance processes need to be thoroughly monitored and evaluated, to prevent such strategic deconstructions of the process.

Closure, on the other hand, is needed to restrict the selection of management options, to guarantee an efficient use of resources, be it financial or the use of time and effort of the participants in the governance process. Closure concerns the part of generating and selecting risk management options, more specifically: Which options are selected for further consideration, and which options are rejected. Closure therefore concerns the product of the deliberation process. It describes the rules of when and how to close a debate, and what level of agreement is to be reached. The quality of the closure process has to meet the following requirements (IRGC 2005: 50; Webler 1995; Widson and Willis 2004):

- Have all arguments been properly treated? Have all truth claims been fairly and accurately tested against commonly agreed standards of validation?
- Has all the relevant evidence, in accordance with the actual state-of-the-art knowledge, been collected and processed?
- Was systematic, experimental, and practical knowledge and expertise adequately included and processed?
- Were all interests and values considered, and was there a major effort to come up with fair and balanced solutions?
- Were all normative judgments made explicit and thoroughly explained? Were normative statements derived from accepted ethical principles or legally prescribed norms?
- Were all efforts undertaken to preserve plurality of lifestyle and individual freedom and to restrict the realm of binding decisions to those areas in which binding rules and norms are essential and necessary to produce the outcome?

If these requirements are met, there is at least a real chance to be able to achieve consensus and a better acceptance of the outcomes of the needed risk assessment options, when facing risk problems with high complexity, uncertainty, and ambiguity. The success of the stakeholder involvement strongly depends on the quality of the process. Consequently, this process has to be specifically designed for the context and characteristics of the corresponding risk. The balance of inclusion and closure is one of the crucial tasks of risk governance.

4.2 Coping with the Plurality of Knowledge and Values

The different social groups enter the governance process with very different preconditions regarding their knowledge about the risk characteristics. In the first chapter it has been set out, that the perception of risks varies greatly among different actor groups. Even among different scientific disciplines, the concepts of risk are highly variable. All the varying types of knowledge and the existing plurality of values have to be taken into consideration, if acceptable outcomes of the risk governance process are aspired. The only possibility to include all these plural knowledge bases and values, are to embed procedures for participation into the governance process.

Depending on the nature of the risk, and the available information about the risk, different levels of public and stakeholder participation seem appropriate to guarantee the quality of the process, if time and effort of the participating groups are regarded as spare resources. In the context of the described risk governance framework, suggestions for the participation of the public and stakeholders have been made depending on the nature of the risk (IRGC 2005: 51f.). Four types of "*discourses*," describing the extent of participation, have been suggested.

In the case of *simple risk problems* with obvious consequences, low remaining uncertainties and no controversial values implied, like many voluntary risks, for example, smoking, it seems not necessary and even inefficient to involve all potentially affected parties to the process of decision making. An "*instrumental discourse*" is proposed to be the adequate strategy to deal with these risks. In this first type of discourse, agency staff, directly affected groups (like product or activity providers and immediately exposed individuals) and enforcement personnel are the relevant actors. It can be expected that the interest of the public into the regulation of these types of risk is very low. However, regular monitoring of the outcomes is important, as the risk might turn out to be more complex, uncertain or ambiguous than characterized by the original assessment.

In case of *complex risk problems* another discourse is needed. An example for complexity-based risk problems are the so-called cocktail effects of combined pesticide residues in food. While the effects of single pesticides are more or less scientifically proven, the cause and effect chains of multiple exposure of different pesticides via multiple exposure routes are highly complex. As complexity is a problem of insufficient knowledge about the coherences of the risk characteristics, which is in itself not solvable, it is more important to produce transparency over the subjective judgments and about the inclusion of knowledge elements, in order to find the best estimates for characterizing the risks under consideration. This "*epistemological discourse*" aims at bringing together the knowledge from the agency staff of different scientific disciplines and other experts from academia, government, industry, or civil society. The principle of inclusion is bringing new or additional knowledge into the process and aims at resolving cognitive conflicts. Appropriate instruments of this discourse are Delphi, Group Delphi, or consensus workshops (Webler et al. 1991; Gregory et al. 2001).

In the case of risk problems due to *high unresolved uncertainty*, the challenges are even higher. The problem here is: How can one judge the severity of a situation when the potential damage and its probability are unknown or highly uncertain? This dilemma concerns the characterization of the risk as well as the evaluation and the design of options for the reduction of the risk. Natural disasters like tsunamis, floods, or earthquakes are, for example, characterized by high uncertainty. In this case, it is no longer sufficient to include experts into the discourse, but policy makers and the main stakeholders should additionally be included, to find consensus on the extra margin of safety in which they would be willing to invest in order to avoid potentially – but uncertain – catastrophic consequences. This type is called *"reflective discourse,"* because it is based on a collective reflection about balancing the possibilities for over- and under-protection. For this type of discourse, round tables, open space forums, negotiated rule-making exercises, mediation or mixed advisory committees are suggested (Amy 1983; Perritt 1986; Rowe and Frewer 2000).

If risk problems are due to *high ambiguity*, the most inclusive strategy is required, as not only the directly affected groups have something to contribute to the debate, but also the indirectly affected groups. If, for example, decisions have to be taken concerning the use or the ban of genetically modified foods and their production, the problem if going far beyond the mere risk problem, but touches also principal values and ethical questions, and questions of lifestyle or future visions. A *"participative discourse"* has to be organized, where competing arguments, beliefs, and values can be openly discussed. This discourse affects the very early step of risk framing and of risk evaluation. The aim of this type of discourse is to resolve conflicting expectations through identifying common values, defining options to allow people to live their own visions of a "good life," to find equitable and just distributions rules for common resources, and to activate institutional means for reaching common welfare so that all can profit from the collective benefits. Means for leading this normative discourse are, for example, citizen panels, citizen juries, consensus conferences, ombudspersons, citizen advisory commissions, etc. (Dienel 1989; Fiorino 1990; Durant and Joss 1995; Armour 1995; Applegate 1998).

In this typology of discourses, it is presupposed, that the categorization of risks into simple, complex, uncertain, and ambiguous is uncontested. But, very often, this turns out to be complicated. Who decides whether a risk issue can be categorized as simple, complex, uncertain, or ambiguous? To resolve this question, a *meta-discourse* is needed, where the decision is taken, where a specific risk is located and in consequence, to which route it is allocated. This discourse is called *"design discourse,"* and is meant to provide stakeholder involvement at this more general level. Allocating the risks to one of the four routes has to be done before assessment starts, but as knowledge and information may change during the governance process, it may be necessary to reorder the risk. A means to carry out this task can be a screening board that should consist of members of the risk and concern assessment team, risk managers, and key stakeholders. Figure 1.6 provides an overview of the described discourses depending on the risk characteristics and the actors included into these discourses. Additionally, it sets out the type of conflict produced through the plurality of knowledge and values and the required remedy to deal with the corresponding risk.

Of course, this scheme is a simplification of real risk problems and is meant to provide an idealized overview for the different requirements related to different risk problems. Under real conditions, risks and their conditions often turn out to be more interdependent among each other and the required measures more depending from unique contexts. This is why actually, the effectiveness of these types of stakeholder involvement are tested in "reality" in a series of very differing risk fields.

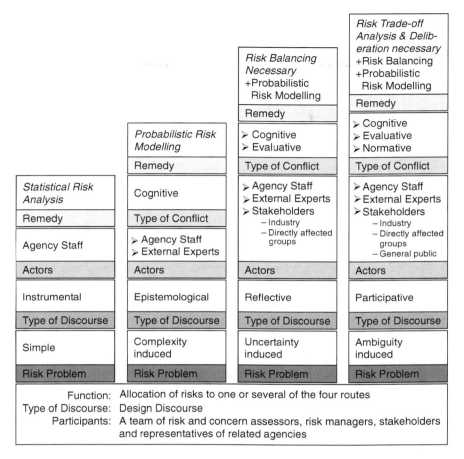

Fig. 1.6 The risk management escalator and stakeholder involvement (IRGC 2005, adapted and modified, p. 280)

4.3   International Challenges When Dealing with Transboundary Risks

It has been foreshadowed in the last paragraph that those responsible for the governance of risks and those affected by risks do normally not face such ideal structures, where they can easily decide which governance routes and measures to take to deal with the problems of complexity, uncertainty, and ambiguity. Often, the risks individuals, companies, regions or countries have to face do not depend on their own choices. Additionally, they often do not only face one risk at a time. For example consumer groups are facing an in-depth discussion about genetically modified food, and a new issue like nanotechnology arises where the public awareness of the risk is at early stage. As a consequence, they have to find strategies to deal with a series of interrelated risks that are often ill-defined or outside of their control (IRGC 2005: 48).

Globalization has contributed to the fact, that *interdependencies* in many cases do not require spatial proximity. For example, diseases through aggressive viruses like the avian flu can easily spread to other regions through single contacts. Another example from the thematic area of the decrease of biodiversity, is the involuntary spread of a risk of invasive species (be it animals or plants). On the one hand, expensive measures are taken to preserve the habitats of endangered species in order to protect them from extinction. But then, on the other hand, it can occur that foreign species are introduced involuntarily via global transports, etc. This species then sometimes displaces the ones that have been tried to preserve. This has, for example, happened in the US Great Lakes region with some species of fish through the invasion of zebra mussels and other species through cargo ships.[8]

These are only two examples for the various challenges when facing the interdependencies produced through transboundary risks. The level of interdependencies adheres to another problem that is typical for global systemic risks: The "goods" (or, as described in this chapter, "what humans value") that are endangered through the risk are often *common goods*, which means that no one can be excluded from its use or profit. Public health is a nonmaterial example for such a common good. The more interdependencies there are within a particular risk situation the smaller is the probability that risk reduction measures are taken. A characteristic of common goods is, that everyone can profit from their use, even if one does not invest in their maintenance. From an individual point of view, a rational actor (be it an individual, a company, a country or any other entity) would act as "*free riders*," i.e., benefit from the use of the good but not contribute to its maintenance. In terms of risk problems, such an actor would not invest in risk reduction measures, while he would profit from the risk reduction measures conducted by other actors. From a collective point of view, each actor would have been better off had all actors invested in the maintenance of the good. The more interdependencies there are, the less an individual actor feels accountable for investing in risk reduction measures.[9] Weak links between the affected parties contribute to this suboptimal behavior. Anthropogenic climate change through the burning of fossil fuels and the production of greenhouse gases is a classical example, but the depletion of biodiversity can also be understood as a free-rider effect of a global dimension.

The global nature of systemic risks and the high level of interdependencies ask for a balanced strategy of consensual, coercive, and incentive-based measures. *Consensual* measures are, for example, international agreements, international standards or gentleman's agreements. *Coercive* measures can be government's regulations and examples for *incentive-based* measures are emission certificates.

---

[8]Cf. for example http://www.greatlakesforever.org/html/trouble/species.html.

[9]A global overview over game theory and the problem of common goods would exceed the scope of this chapter. For a more formal theoretic treatment of the problem cf. Kunreuther and Heal (2003); for the free-rider problematic cf. Cornes and Sandler (1996). The "tragedy of the commons" is classically described in Hardin (1968).

Which kinds of measures are appropriate depends on the degree of decentralization, the political culture, and the associated regulatory styles.

One possible solution for the management of the described effects due to inter-dependencies and the resulting individual rationality and losses in accountability due to weak links might be *public private partnerships (PPP)* (IRGC 2005: 48). PPPs can be defined as an agreement or co-operation between the public and the private sector and is often understood as a variation of privatization. It is very often applied for the development and maintenance of infrastructure measures. PPPs seem to be particularly adequate if the risks to be dealt with are associated with competing interpretations (ambiguities) as to what type of co-operation is required between different scientific communities and risk management agencies in order to be able to deal with various types of knowledge and competing values.

A possible way to structure such partnerships is to have government standards and regulations coupled with third party inspections and insurance in order to enforce these measures. It is thus a management-based regulatory strategy that will not only encourage the addressees of the regulation, often the corporate sector, to reduce their risks from, for example, accidents and disasters. It forces the actors of the private sector to do their own planning as to how they can meet the given standards or regulations and so shift the decision making from the government regulatory authority to the private company. This might be of advantage as the companies can choose the means and measures that most fit for the purpose within their specific environmental context, and may lead to a optimized allocation of resources compared to more top-down forms of regulation.

In the case of risks resulting from large plants (be it power plants or chemical sites, etc.), for example, the combination of third party inspections together with private insurance can turn out to be a powerful combination of public over-sight and market mechanisms and can thus convince many companies of the advantages of implementing the necessary measures to make their plants safer and encourage the remaining ones to comply with the regulation to avoid being caught and prosecuted.

Consequently, PPPs are an effective means for the internalization of external effects, i.e., the problem of weak links produced through a high level of interdepen-dencies are strengthened by accounting responsibility for the consequences of risk-producing actions to single actors.

5 Premises for Successful Risk Governance

5.1 Organizational Capacity to Deal with Risks

In Chap. 2, a short overview over the different phases and aspects of risk gover-nance and their interrelations has been given. This chapter aims for answering the question which specific steps are needed to handle systemic risks. In Chap. 3 some core challenges, like varying values and cultural settings as well as interdependencies, for the governance process have been set out. But one important question has been left open so far: Do the governing actors have the capability to deal with systemic risks as proposed above? If so, what are the prerequisites to fulfill their proposed roles? How has the interplay between the different actors to be designed?

It is certainly idealistic to assume that societies, when they face new and emerging risks, have developed the institutional and organizational capacity that is needed to perform the tasks described in the governance framework. The realities of the political context can be exemplified for the very first step in the governance cycle, the process of risk framing (IRGC 2005: 58f.): Bringing specific risk issues on the political agenda and consequently to the media as well, is a common means to wield power or to collect votes during election campaigns. In this manner, it influences the governance process from the beginning. Public dissent due to varying risk perceptions or media hypes in the context of a certain risk are often used to push individual interests (of political parties, for example) (Shubik 1991). Such influences together with the potential of mobilization of the wider public, can lead into the playing up of some risks while other risks might be concealed or downplayed due to individual motivations.

As a consequence, many political systems have reacted by establishing independent risk assessment and sometimes management agencies, in order to prevent such exertions of influence. The establishment of numerous national and the European food standards agencies is the most cited example for the institutional restructuring of risk governance. In the mid-1990s, when the transmission of the cattle disease Bovine Spongiform Encephalopathy (BSE) to man was discovered in the UK, in the shape of a new variant of the Creutzfeldt–Jakob disease (vCJD), a policy of reassurance and inadequate scientific attention led to the biggest food scandal in the twentieth century, as measured its consequences: "no debate inside the European Union (EU) was more influential to everyday life than BSE; no other food scandal had a bigger impact on the public discourse of eating habits or regarding questioning conventional farming practices" (Dressel 2002: 60). This scandal led to several institutional changes within the EU and was the motor of the establishment of the European Food Standards Agency (EFSA) and several national food standards agencies or independent risk assessment agencies, like the Federal Institute for Risk Assessment (Bundesinstitut für Risikobewertung, BfR) in Germany, and the Food Standards Agency (FSA) in the UK (Dressel 2002; Dreyer et al. 2006).

5.2 Prerequisites of Good Governance

What lessons can be leant from this and other failings in the governance of risks? First, it is important to make sure that the governance process is based on the best available knowledge and practice. Second, institutions and organizations have to be strengthened so that they are empowered and have the resources to perform their tasks in the most possible effective, efficient, and fair manner (IRGC 2005: 58).[10]

---

[10]Cf. also the next chapter on the prerequisites of good governance.

To make sure, that the responsible institutions and organizations are able to act in that way, three analytic categories can be used to assess institutional capacity (ibid. Paquet 2001):

- The knowledge bases and the structural conditions for effective risk management build the *assets* of the governance institutions. This category includes *rules, norms and regulations*, available *resources* (financial as well as infrastructure, but also access to and processing of information), *competencies and knowledge* (in terms of education and training), and the level of *organizational integration* of the aforementioned types. Organizational integration can be understood as the prerequisite for the effective use of the other types and in a mathematical sense as a multiplying factor.
- *Skills* describe the quality of the institutional and human performance in exploring, anticipating and dealing with existing and emerging risks, here understood as often unpredictable external conditions. They should enable political, economic, and civic actors to use effectively, and enhance the impact of, the described assets. Skills include *flexibility* (i.e., openness to make use of new ways in dynamic situations), *vision* (making use of new methods that are normally used in other contexts, e.g., foresight, scenario planning, etc.) and *directivity* (expand the risk context into a reframing of the whole perception if the way of life and thus driving change that impacts on the outside world instead of restricting oneself to the prevention or mitigation of external effects).
- The framework, in which assets and skills can be exploited for the development and exploitation of successful risk governance policies is built by the last category, the *capabilities*. Consequently, they build the structure and include *relations* (manage the inclusion through linking users and sources of knowledge; those carrying the authority and those bearing the risk), *networks* (constitute close co-operative structures between self-organization and hierarchy between and among groups of principally equal actors) and *regimes* (establish the rules, the frameworks and are formed through the two types above).

As a prerequisite for the building and functioning of these three categories, *risk education and training* have to be seen as fundamental resources for making use of the "human capital" in order to handle global, emerging and systemic risks. Such education and training measures should aim at a broad and multidisciplinary knowledge base instead of specialized in-depth knowledge, to be able to deal with the challenges of interdependencies, complexity, uncertainty and ambiguities. The often predominating technical focus in scientific education therefore needs to be expanded to health, safety, and environmental aspects, i.e., enabling to actors to take up a "bird's eye perspective" (IRGC 2005: 61).

5.3 Principles of the Governance of Systemic Risks

The term risk governance, as it has been set out in this chapter, denotes not only the governmental actions taken toward the mitigation or prevention of risk consequences, but the whole interplay of all relevant actors – and all actions that are undertaken to handle risks. The integration of so many different views and interests, values and norms creates a very complex structure, which is difficult to comprehend for the public and great parts of the affected groups as well. In order to ensure the functioning of such a complex and interdependent formation, where direct links between the different parties and tasks are often absent or too weak due to international or global dimensions of the risk problems, some general principles have to be set up to support a governance process with outcomes that are accepted or at least tolerated.

On the European level, the Commission has carried out this task in order to strengthen its democratic structures while working on solutions to the major problems confronting European societies, like demographic changes, health risks like smoking, food safety scares, crime, and unemployment. Anyway, interest as well as confidence and trust into the work of the European institutions have decreased during the last years. At the same time, the Europeans expect the governments and the European Union to take the lead in reducing risks which emerge in the context of globalization, growth of the population, and the economic development. This is particularly true for the handling of international systemic risks. For the improvement of people's trust and confidence into the performance of the European institutions, the European Commission has worked out a White paper (European Commission 2001), in which a number of principles of good governance are described, which should help them to carry out the task needed for the governance of, for example, systemic risks (ibid. 10f.):

- *Openness*: The institutions responsible for the assessment and management of risks should work in an open and transparent manner. This means they should actively communicate to the affected and interested parties and the stakeholders about their tasks, lay open their structures and what and how decisions are taken. This includes the use of a language that is accessible and understandable for the general public, in order to improve the confidence in complex structures and decisions.
- *Participation*: Inclusion of stakeholders and the affected and interested public is set as a crucial task of risk governance. Acceptance in decisions about the handling of risks, and confidence in the outcomes of governance processes depend on the inclusion of the interested parties throughout the whole governance chain.
- *Accountability*: Roles and responsibilities of the different actors in the governance process have to be made clear. From a European point of view, it has to be made clear, which institutions carry out which tasks and where they have responsibility on national and international level. Additionally, the specific tasks of the involved parties in the different stages of the risk governance process have to be made clear.
- *Effectiveness*: Risk governance policies have to be effective and timely, have to deliver what is needed on the basis of clear objectives, an evaluation of future

impact and, where available, of past experience. Time and effort have to be treated as spare resources. Measures have to follow the principles of proportionality and appropriateness.

- *Coherence*: Policies and actions have to be coherent and easily understood. As the range and complexity of institutions is constantly growing, interdependencies between different sectors are increasing, regional and local authorities are increasingly involved in European policies, etc. These tendencies require political leadership, including a strong responsibility from institutional side, to guarantee consistent procedures within this complexity.
- *Proportionality and subsidiarity*: Throughout the whole governance process, the choice of the level at which the action is taken (from European to local level) and the selection of the instruments used must be considered in the proportion to the objectives pursued.

The compliance with these principles poses high challenges to those who design and those who carry out the different steps of the risk governance process. It is possible that the adherence to one principle complicates the adherence to another. So, for example, more inclusion and participation might be seen as ineffective by some actors. So the main challenge is to find a balance, i.e., to decide which level of participation is really necessary, which decision have to be taken on European level, and which on national or regional level, and who decides if the chosen measures are proportionate to the achievable objectives.

6   Chapter Summary

This first introductory chapter was meant to give the reader an overview over current risk science, crucial elements of risk governance and an impression of where the new developments and approaches are leading us when thinking about risks and how to deal with them.

It was suggested to look at risk governance not as a linear process of risk analysis, risk management, and risk communication of ready-made results, as taught us the traditional approach of risk analysis, but as a circular process, including public risk perceptions, values, and concerns. A framework, developed by the IRGC, has been presented which takes into account these "human factors" of risk, and which understands risk governance as a cycle with the possibility of feedback loops and proposes a set of specific discourses for stakeholder involvement according to the risk characteristics (simple, complex, uncertain, or ambiguous).

This chapter should have equipped the reader with the needed knowledge of approaches, frameworks, models, and tendencies to be able to better understand the commonalities and differences of the following chapters, dealing with such different risk fields as genetically modified foods, food risks from dioxins and the loss of biodiversity. So, when reading these more case study type chapters, keep the presented analytic structure in mind.

Glossary

ALARP – principle: A term often used in the milieu of safety-critical and high-integrity systems. The ALARP principle says that the residual risk shall be "*As Low As Reasonably Practicable.*"

Ambiguity: Giving rise to several meaningful and legitimate interpretations of accepted risk assessment results.

Closure: Describes the restriction of the selection of management options, to guarantee an efficient use of resources, be it financial or the use of time and effort of the participants in the governance process.

Complexity: Refers to the difficulty of identifying and quantifying causal links between a multitude of potential causal agents and specific observed effects.

Governance: Describes the structures and processes of collective decision making, including governmental as well as nongovernmental actors.

Hazard: A source of potential harm or a situation with the potential to cause loss.

Inclusion: Describes the question of what and whom to include into the governance process, not only into the decision making, but into the whole process from framing the problem, generating options and evaluating them to coming to a joint conclusion.

Persistence: Describes the timescale, how long a damage lasts.

Risk: An uncertain consequence of an event or an activity with respect to something that humans value. The judgment, if these consequences are seen as positive or negative depends on the values that people associate with them.

Risk analysis: Used by a number of organizations dealing with risk as a collective term for risk assessment, risk management, and risk communication.

Risk assessment: The task of identifying and exploring, preferably in quantified terms, intensities and likelihood of the (negative) consequences of a risk.

Risk governance: Includes totality of actors, rules, conventions, processes, and mechanisms concerned with how risk information is collected, analyzed, and communicated and how management decisions are taken.

Risk management: The creation and evaluation of options for initiating or changing human activities or structures with the objective of increasing the net benefit of human society and preventing harm to humans and what they value.

Social amplification of risk: Describes "an overestimation or underestimation of the seriousness of a risk caused by public concern about the risk or an activity contributing to the risk" (IRGC 2005: 81).

Systemic risks: Risks that affect in complete the various systems on which society depends, i.e., health, transport, energy, telecommunications, etc. These risks are at the crossroads between natural events (which can be partially altered and increased by human actions), economic, social, and technological developments and policy-driven actions.

Ubiquity: Geographical dispersion of a damage.

Uncertainty: A state of knowledge in which, although the factors influencing the issues are identified, the likelihood of any harmful effect or these effects themselves, cannot be precisely described.

References

Amy DJ (1983) Environmental Mediation: An Alternative Approach to Policy Stablemates. *Policy Sciences* 15:345–365

Applegate J (1998) Beyond the Usual Suspects: The Use of Citizens Advisory Boards in Environmental Decisionmaking. *Indiana Law Journal* 73:903–957

Armour A (1995) The Citizen's Jury Model of Public Participation. In: Renn O, Webler T, Wiedemann P (eds) *Fairness and Competence in Citizen Participation. Evaluating New Models for Environmental Discourse*. Kluwer, Dordrecht & Boston, pp 175–188

Beck U (1986) *Risikogesellschaft. Auf dem Weg in eine andere Moderne*. Edition Suhrkamp, Frankfurt am Main

Brehmer B (1987) The Psychology of Risks. In: Singleton WT, Howden J (eds.) *Risk and Decisions*. Wiley, New York, pp 25–39

Cornes R, Sandler T (1996) *The Theory of Externalities, Public Goods and Club Goods* 2nd edn. Cambridge University Press, Cambridge

Dienel PC (1989) Contributing to Social Decision Making methodology: Citizen Reports on Technological Projects. In: Vlek C, Cvetkovich G (eds) *Social Decision Methodology for Technological Projects*. Kluwer, Dordrecht & Boston, pp 133–151

Dressel K (2002) *Systemic Risk: A new challenge for risk management. The case of BSE*. Report to the OECD. Sine-Institute, München

Dreyer M, Renn O, Borkhart K, Ortleb J (2006) Institutional Re-Arrangements in European Food Safety Governance: A Comparative Analysis. In: Vos E, Wendler F (eds) *Food Safety Regulation in Europe: A Comparative Analysis*. (Series: Ius Commune). Intersentia Publishing, Antwerpen, pp 9–64

Dreyer et al. (2009) Summary: Key Features of the General Framework. In: Dreyer M and Renn O (eds) *Food Safety Governance. Integrating Science, Precaution and Public Involvement*. Springer: Heidelberg and New York, pp 159–166

Drottz-Sjöberg BM (1991) *Perception of Risk. Studies of Risk Attitudes, Perceptions, and Definitions*. Center for Risk Research, Stockholm

Durant J, Joss S (1995) *Public Participation in Science*. Science Museum, London

European Commission (2001) *European Governance. A White Paper*. EU, Brussels

Fiorino DJ (1990) Citizen Participation and Environmental Risk: A Survey of Institutional Mechanisms. *Science, Technology & Human Values* 15(2):226–243

Fischhoff B, Slovic P, Lichtenstein S, Read S, Combs B (1978) How Safe is Safe Enough? A Psychometric Study of Attitudes Toward Technological Risks and Benefits. *Policy Sciences* 9:127–152

Fischhoff, B., Watson, S. R. and Hope, C. (1984) 'Defining risk', Policy Sciences, vol 17, pp123–129

Fischhoff B, Bostrom A, Jacobs-Quadrel M (1993) Risk perception and communication. *Annual Review of Public Health* 14:182–203.

Gregory R, McDaniels T, Fields D (2001) Decision Aiding, Not Dispute Resolution: A New Perspective for Environmental Negotiation. *Journal of Policy Analysis and Management* 20 (3):415–432

Guerin B (1991) Psychological perspectives on risk perception and response. In: Handmer J, Dutton B, Guerin B, Smithson M (eds) *New perspectives on uncertainty and risk*. ANU/CRES, Canberra, pp 79–86

Hardin G (1968) The Tragedy of the Commons. *Science* 162:1243–1248

IRGC (2005) *White Paper on Risk Governance. Towards an Integrative Approach*. International Risk Governance Council, Geneva

Jasanoff S (1986) *Risk Management and Political Culture*. Russell Sage Foundation, New York

Jasanoff S (2004) Ordering Knowledge, Ordering Society. In: Jasanoff S (ed) *States of Knowledge: The Co-Production of Science and Social Order*. Routledge, London, pp 31–54

Jaeger C, Renn O, Rosa E, Webler T (2001) *Risk, Uncertainty and Rational Action*. Earthscan, London

Jungermann H, Slovic P (1993) Die Psychologie der Kognition und Evaluation von Risiko. In: Bechmann G (ed) *Risiko und Gesellschaft: Grundlagen und Ergebnisse interdisziplinärer Risikoforschung*. Westdeutscher Verlag, Opladen, pp 167–207

Kates RW, Hohenemser C, Kasperson J (1985) *Perilous Progress: Managing the Hazards of Technology*. Westview Press, Boulder

Klinke A, Renn O (2006) Systemic Risks as Challenge for Policy Making in Risk Governance. Forum Qualitative Sozialforschung/Forum: Qualitative Social Research [Online Journal] 7(1), Art. 33. http://www.qualitative-research.net/fqs-texte/1-06/06-1-33-e.htm. Accessed 13 March 2006

Kunreuther H, Heal G (2003) Interdependent Security. *Journal of Risk and Uncertainty*, Special Issue on Terrorist Risks 26(2/3):231–249

Lasswell, H. D. (1948) 'The structure and function of communication in society', in L. Brison (ed) *The Communication of Ideas*, New York, NY, pp32–51

Leiss W (1996) "Three Phases in Risk Communication Practice." Annals of the American Academy of Political and Social Science, Special Issue 545:85–94.

Löfstedt RE, Vogel D (2001) The Changing Character of Regulation. A Comparison of Europe and the United States. Risk Analysis 21(3):393–402

Morgan MG, Fischhoff B, Bostrom A, Lave L, Atman C (1992) Communicating Risk to the Public. *Environmental Science and Technology* 26(11):2049–2056

Nye JS, Donahue J (eds) (2000) *Governance in a Globalising World*. Brookings Institution, Washington

OECD (2002) *Guidance Document on Risk Communication for Chemical Risk Management*. OECD, Paris

OECD (2003) *Emerging systemic risks*. Final report to the OECD futures project. OECD, Paris

Paquet G (2001) The New Governance, Subsidiarity, and the Strategic State. In: OEDC (ed) *Governance in the 21ˢᵗ Century*. OECD, Paris, pp 183–215

Perritt HH (1986) Negotiated Rulemaking in Practice. *Journal of Policy Analysis and Management* 5:482–495

Perrow C (1992) *Normale Katastrophen. Die unvermeidbaren Risiken der Großtechnik*. Campus, Frankfurt am Main

Pidgeon NF (1998) Risk Assessment, Risk Values and the Social Science Programme: Why We Do Need Risk Perception Research. *Reliability Engineering and System Safety* 59:5–15

Pidgeon NF, Hood CC, Jones DKC, Turner BA, Gibson R (1992) Risk Perception. In: Royal Society Study Group (ed) *Risk Analysis, Perception and Management*. The Royal Society, London, pp 89–134

Renn O (2004) Perception of Risks. *The Geneva Papers on Risk and Insurance* 29(1):102–114

Renn O (1992) Concepts of Risk: A classification. In: Krimsky S, Golding D (eds) *Social Theories of Risk*. Praeger, Westport/London, pp 53–79

Renn O (1990) Risk Perception and Risk Management: A Review. *Risk Abstracts* 7(1):1–9

Renn O (1986) Decision Analytic Tools for Resolving Uncertainty in the Energy Debate. *Nuclear Engineering and Design* 93(2&3):167–180

Renn O, Klinke A (2004) Systemic Risks: A New Challenge for Risk Management. *EMBO Reports*, Special Issue 5:541–546

Renn O, Kastenholz H (2000) Risk Communication for Chemical Risk Management. In: Bundesinstitut für gesundheitlichen Verbraucherschutz und Veterinärmedizin (ed) An OECD Background Paper for the OECD-Workshop Berlin

Rohrmann B (1995) *Risk Perception Research. Review and Documentation. Arbeiten* zur Risiko-Kommunikation 48. Forschungszentrum Jülich, Jülich

Rohrmann B, Renn O (2000) Risk Perception Research – An Introduction. In: Renn O, Rohrmann B (eds) *Cross-Cultural Risk Perception. A Survey of Empirical Studies.* Kluwer, Dordrecht & Boston, pp 11–54

Rosenau JN (1992) Governance, Order and Change in World Politics. In: Rosenau JN, Czempiel EO (eds) *Governance without Government, Order and Change in World Politics.* Cambridge University Press, Cambridge, pp 1–29

Rowe G, Frewer L (2000) Public Participation Methods: An Evaluative Review of the Literature. *Science, Technology and Human Values* 25:3–29

Schramm W (1954) "How Communication Works" In: Schramm W (ed) The Process and Effects of Communication. University of Illinois Press, Illinois, pp 3–26

Shannon, C. E. and Weaver, W. (1949) The Mathematical Theory of Communication, University of Illinois Press, Urbana, IL

Shubik M (1991) Risk, Society, Politicians, and People. In: Shubik M (ed) *Risk, Organizations, and Society.* Kluwer, Dordrecht & Boston, pp 7–30

Sjöberg L, Drottz-Sjöberg B-M (1994) *Risk perception of nuclear waste: experts and the public* (Rhizikon: Risk Research Report 16). Center for Risk Research, Stockholm School of Economics

Slovic P (1987) Perception of Risk. *Science* 236:280–285

Slovic P (1992) Perception of Risk: Reflections on the Psychometric Paradigm. In: Krimsky S, Golding D (eds) *Social Theories of Risk.* Praeger, Westport and London, pp 153–178

Stern PC, Fineberg V (1996) *Understanding Risk: Informing Decisions in a Democratic Society.* National Research Council, Committee on Risk Characterisation. National Academy Press, Washington DC

Trichopoulou et al. (2000) *European Policy on Food Safety,* report to the European Parliament's Scientific and Technological Options Assessment Programme (STOA), PE number: 292.026/ Fin.St, September 2000: 68

Trustnet (1999) *A New Perspective on Risk Governance. Document of the Trustnet Network.* EU, Paris. http://www.trustnetgovernance.com. Accessed 14 May 2006

US National Research Council (1989) Improving Risk Communication, National Academy Press, Washington, DC

van Asselt, M. B. A. (2000) Perspectives on Uncertainty and Risk, Kluwer, Dordrecht and Boston, MA

Von Gleich A (1999) Ökologische Kriterien der Technik- und Stoffbewertung: Integration des Vorsorgeprinzips, Teil II: Kriterien zur Charakterisierung von Techniken und Stoffen. *Umweltwissenschaften und Schadstoff-Forschung - Zeitschrift für Umweltchemie und Ökotoxikologie* 11(1):21–32

Von Gleich A (2003) Mit Nanotechnologie zur Nachhaltigkeit? Charakterisierung der Technologie und leitbildorientierte Gestaltung als Auswege aus dem Prognosedilemma der Technikbewertung. In: Böschen S, Lerf A, Schneider M (eds) *Über die Anerkennung von und den Umgang mit Nichtwissen.* Edition sigma, Berlin

WBGU (2000) *World in transition. Strategies for managing global environmental risks.* Annual Report 1998. Springer, Berlin German Advisory Council on Global Change

Webler T (1995) Right Discourse in Citizen Participation. An Evaluative Yardstick. In: Renn O, Webler T, Wiedemann P (eds) *Fairness and Competence in Citizen Participation. Evaluating New Models for Environmental Discourse.* Kluwer, Dordrecht & Boston, pp 35–86

Webler T (1999) The Craft and Theory of Public Participation: A Dialectical Process. *Risk Research* 2(1):55–71

Webler T, Levine D, Rakel H, Renn O (1991) The Group Delphi: A Novel Attempt at Reducing Uncertainty. *Technological Forecasting and Social Change* 39:253–263

Widson J, Willis R (2004) *See-Through Science. Why Public Engagement Needs to Move Upstream.* Demos, London

Wolf KD (2002) Contextualizing Normative Standards for Legitimate Governance Beyond the State. In: Grote JR, Gbikpi B (eds) *Participatory Governance. Political and Societal Implications.* Leske und Budrich, Opladen, pp 35–50

Wynne B (2002) Risk and Environment as Legitimatory Discourses of Technology: Reflexivity Inside Out? *Current Sociology* 50(30):459–477

Chapter 2
Biodiversity at Risk

Juliette Young and Allan Watt

1 Why Are We Writing About This Topic?

Biological diversity, or biodiversity, is broadly defined as the variety of life on Earth and the natural patterns it forms. Almost two million species have now been identified and the actual number of species in the world is estimated to be between 10 and 30 million (IUCN 2006). This enormous biodiversity is an essential provider of ecosystem goods and services. However, despite the important role biodiversity plays in our lives, all species that together comprise biodiversity face risk.

It is important, however, to differentiate between different types of risk to biodiversity. The science of ecology arose from the study of the way that animals and plants respond to *natural risks* in their environment and although ecologists have often quantified the enormous mortality caused by many factors, they have also identified the traits that organisms have evolved to cope with the risks that they are exposed to. Although all species face natural risks and are adapted to them, biodiversity is increasingly facing *new risks*, mainly caused by human activities, that species cannot respond to quickly enough and therefore threaten their survival.

Perhaps the first example of the widespread realization that biodiversity faced risks that it has not evolved the ability to respond to was the organochlorine insecticide DDT. This and related insecticides were developed in the 1930s and used to control agricultural pests and insect-borne diseases such as malaria. However, they were also found to be responsible for the decline of predatory birds because they are slow to break down in the environment and accumulate in predators as they consume contaminated prey. In 1962, Rachel Carson published *Silent Spring*, highlighting the risk of pesticides. Although DDT and related pesticides are particularly toxic to fish and other aquatic species, most attention was focussed on their impact on birds of prey such as the bald eagle *Haliaeetus leucocephalus* (Bowerman et al. 1998) and the sparrowhawk *Accipiter nisus* (Newton and Wyllie 1992).

J. Young (✉)
NERC Centre for Ecology and Hydrology, Bush Estate,
Penicuik EH26 0QB, United Kingdom
e-mail: jyo@ceh.ac.uk

P. Pechan et al., *Safe or Not Safe: Deciding What Risks to Accept in Our Environment and Food*, DOI 10.1007/978-1-4419-7868-4_2, © Springer Science+Business Media, LLC 2011

These pesticides and their breakdown products are a major risk because they affect eggshell thickness and hatching success. This led to restrictions in their use and a greater awareness of the risk of novel chemicals in the environment.

Pesticides and other chemicals still pose major risks to biodiversity; organochlorine insecticides are still responsible for mortality in birds such as the white-faced ibis *Plegadis chihi* in the USA (King et al. 2003). However, recent global assessments list many other risks.

The five most important risks to biodiversity at the global scale were identified by the Millennium Ecosystem Assessment (2005) as:

– Habitat change, including habitat loss and fragmentation
– Climate change
– Invasive species
– Overexploitation and persecution
– Pollution, for example, nitrogen deposition and acid rain

The Millennium Ecosystem Assessment also estimated the degree of importance of different risks, or "direct drivers" of change as they named them, in different ecosystems, and their trends (Fig. 2.1). We will discuss several of these risks to biodiversity in this chapter.

These risks affect the survival of many species: about 800 are known to have become extinct since 1600 (Baillie et al. 2004). IUCN, the World Conservation Union, provides regular estimates of the number of globally threatened species. Its 2006 Red List puts the number of known threatened species at 16,119 (IUCN 2006). IUCN's estimates are based on well-known groups of species such as birds; the total number of threatened species is likely to be much larger. Organizations such as IUCN, World Wide Fund for Nature (WWF), Conservation International and the United Nations Environment Programme (UNEP) have been warning of the increasing rate of extinctions for many years and researchers have added to these concerns with many scientists referring to a "biodiversity crisis" (e.g., Koh et al. 2004).

The current rate of species extinction is estimated to be 1,000–10,000 times higher than the estimated "natural rate" of one species out of every million each year (Singh 2002).

This rapid loss of species and habitats is happening throughout the world. Perhaps the most worrying is the current rate of extinction in tropical areas of the world, where most species occur. However, species extinction is also happening in other parts of the world such as Europe, where an estimated 30% of all European plant species are endangered, while 40% of mammal species are now considered to be under threat, especially predatory species such as the brown bear *Ursus arctos*, the lynx *Lynx lynx* and the otter *Lutra lutra* (Stanners and Bourdeau 1995).

With the United Nations predicting a world population of nine billion people in 50 years time, the future pressures on and risks to biodiversity and its services are likely to be significantly more considerable than they are at present and, as such, require increased efforts to better understand and manage such risks. Another reason for discussing risk to biodiversity in the context of the direct risk caused by human activities is that other chapters in this book focus on the opposite – environmental

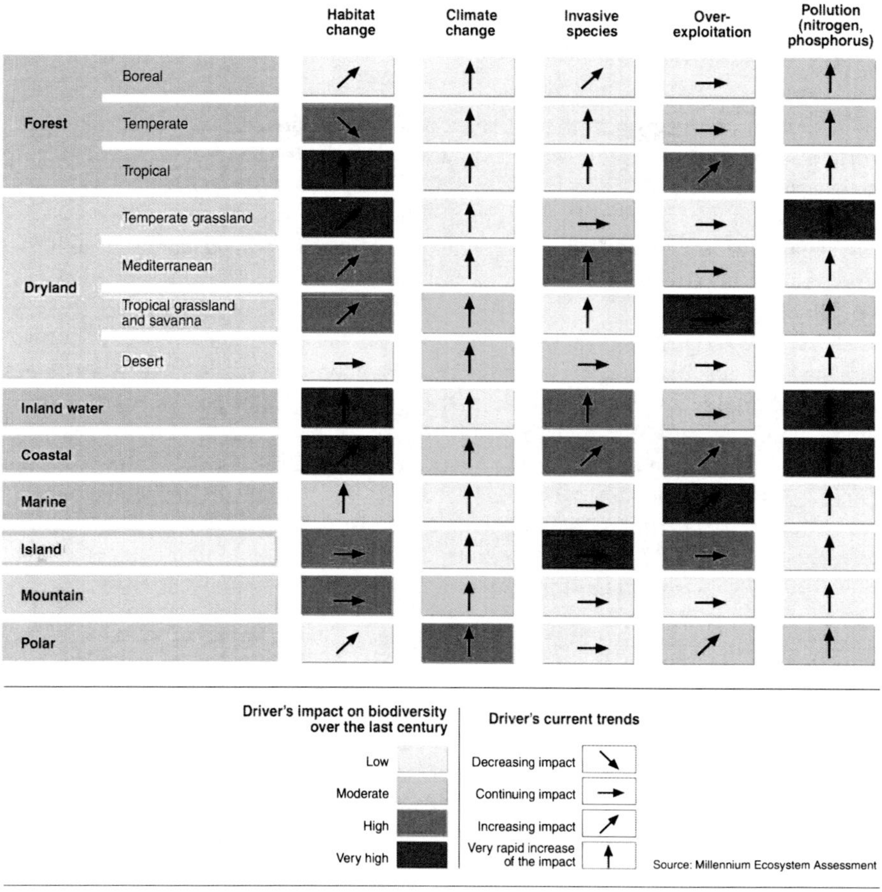

Fig. 2.1 Main direct drivers of change in biodiversity and ecosystems identified by the Millennium Ecosystem Assessment (2005). The impact of each driver on biodiversity in each ecosystem type over the last 50–100 years is indicated by the color of the cells, high impact, for example, indicating significant change to biodiversity in the relevant biome due to the particular driver. Trend is indicated by *arrows*: increasing trends in impact are represented by *diagonal* and *vertical arrows* and continuation of the current levels of impact are shown by *horizontal arrows*

risks to humans. These contrasting perspectives lead us naturally to a recurring theme in considering risk to biodiversity, that of conflict between the conservation of biodiversity and human activities such as hunting, fishing, and a range of agricultural practices. The latter include the use of genetically modified crops, the topic of Chap. 4. That chapter considers the risk of such crops to human health. The possibility that genetically modified organisms also pose a risk to biodiversity has also been considered (Hails 2005). The aim of this chapter is therefore to provide an overview of biodiversity risks before presenting a range of existing options to assess and manage biodiversity risks.

2   Why Is Biodiversity Important for the World and for Everyday Life?

There are no simple answers to the question of why biodiversity is important for the world and everyday life and why we should take steps to conserve biodiversity. On the one hand, one can argue that all species have the inherent right to exist and therefore should be conserved at all cost. On the other hand, one species more or less, what does it matter in the grand scheme of things? To make some sense of the issue, we can unpick the different values of biodiversity, namely direct and indirect use values, and the way in which these values impact on everyday life.

Direct use values of biodiversity include all the goods or products used by humans. Perhaps the most important is the provision of food. All our food comes from other organisms. It is estimated that over 7,000 species of plants and several hundreds of species of animals have been used, at one time or another, for food consumption (Millennium Ecosystem Assessment 2005). Although nowadays only about 20 species of plant account for 90% of all vegetable food consumption, it is essential to bear in mind that biodiversity constitutes a valuable and essential resource with over 35,000 edible plants available for exploitation (Leakey and Lewin 1995). Therefore, the potential direct use value of biodiversity in terms of food provision is huge.

Another direct use value is the provision of medicine. Compounds extracted from certain species are used to treat diseases affecting humans. The most famous is probably aspirin, derived from a constituent of meadowsweet *Filipendula ulmaria*. Other compounds include digitalin from foxgloves such as *Digitalis purpurea*, prescribed for patients diagnosed with heart failure; quinine, from the cinchona tree *Cinchona officinalis* is used to combat malaria; and paclitaxel, a compound found in the bark of the Pacific yew tree *Taxus brevifolia* is effective in treating patients with lung, ovarian, and breast cancer.

In addition to these benefits, biodiversity also provides fuel (e.g., through timber and coal), shelter (timber and other forest products used as building materials and for shelter) and fibers (wool and cotton, for example).

As well as these more tangible benefits, biodiversity has a wealth of indirect use values, providing humans and other species with essential ecosystem services such as nutrient cycling, pollination, and regulation of the atmosphere and climate.

All species are supported by the interactions among other species and ecosystems, each providing an ecological value to one another. Producers (plants) get energy from the sun captured through photosynthesis, a process whereby glucose, carbon dioxide, and water are all synthesized by plants creating oxygen as a waste product. Nearly all life depends on this essential process. These same producers gain inorganic nutrients such as carbon, nitrogen, or phosphorus from the atmosphere, water, or soils to produce living biomass. Producers form the food base for primary consumer species (herbivores) who, in turn, provide the food base for secondary consumer species (carnivores). Producers and consumers all produce dead tissue and waste products. Decomposers transform these waste products into living biomass, which, in turn, forms the food base for consumers. Finally, decomposers and consumers produce inorganic nutrients by mineralization, completing the cycling of nutrients between organic and inorganic forms. This process illustrates that if species are lost, the ecosystem might become less resilient and less productive.

Biodiversity also provides a number of regulating processes such as climate and flood control, as well as pest, pollution, and disease control. Biodiversity also plays an essential role in pollination, with thousands of species of bees, birds, bats, wasps, and flies responsible for the pollination of flowering plants. The most important pollinators are bees, which are responsible for the pollination of some 73% of the world's crops (Roubik 1995). Of the approximately 240,000 species of flowering plants with known pollinators, nearly 220,000 are pollinated by animals. Pollination has important repercussions for humans in terms of food production, as an estimated two-thirds of the world's 3,000 species of agricultural crops require animals for pollination. The annual value of pollination in the worldwide is estimated to be worth $65–70 billion (Pimentel et al. 1997).

Biodiversity is important for climate regulation, with plant functional diversity and habitats influencing the sequestration of carbon, evapotranspiration, temperature, and fire regime (Millennium Ecosystem Assessment 2005). Forests affect rainfall patterns through transpiration losses and protect the watershed of vast areas. Deforestation can therefore affect the amount and distribution of rainfall locally. Deforestation also results in erosion and loss of soil, which in turn could lead to flooding. Deforestation through burning also impacts on climate, due to the decrease in the amount of carbon dioxide taken in by plants through photosynthesis and by releasing vast quantities of carbon dioxide into the atmosphere. Both these processes lead to a global increase of carbon dioxide in the atmosphere, one of the main greenhouse gases responsible for global warming.

All these ecosystem services have been the subject of a number of studies, one of which estimated the *direct* economic value of these services to be in the region of US$33 trillion every year (Costanza et al. 1997).

Finally, biodiversity has important indirect use values including cultural and spiritual values, such as esthetic, educational, and recreational values. Biodiversity is often closely linked to cultural identity. Animals are often referred to in religious texts and worshipped as animal deities. In Greek mythology a myriad of half-human half-animal gods exist including Pan, depicted with the hindquarters, legs, and horns of a goat and Poseidon, depicted with a fish tail. In Hinduism, one animal deity is Ganesh, a man with a one-tusked elephant head. Outside of formal religion, many people feel connected to species. Edward O. Wilson (1984) referred to the connections that human beings subconsciously seek with the rest of life as biophilia. This hypothesis explains to a certain degree why people empathize with other species, care for animals, visit zoos, national parks, and aquaria, and often like to grow plants and flowers.

The Millennium Ecosystem Assessment (2005) report describes in greater detail the goods and services touched on in this section. They outline the status of provisioning services such as food, fresh water, wood and fuel; regulating services such as climate regulation, flood regulation, and water purification; supporting services such as primary production, soil formation, and nutrient cycling; and cultural services such as esthetic and recreational services (see Fig. 2.2).

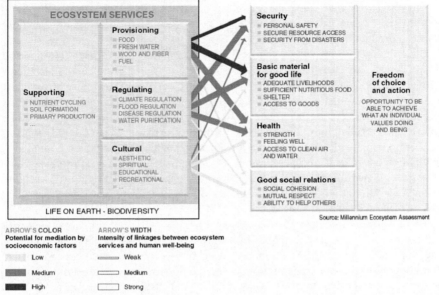

Fig. 2.2 Linkages between ecosystem services and human well-being. *Source*: The Millennium Ecosystem Assessment (2005)

In summary, although we may consider risk to biodiversity as a separate topic to risk to humans, biodiversity, as pointed out above, provides human society with food, fuel, shelter, and many other goods and services that are essential for our survival. Many species do not, of course, contribute directly to human survival. However, some of these species are esthetically or culturally valued. Others possess the genetic material for the crops and drugs of the future. We probably underestimate the value of biodiversity: plants and animals provide the food we eat; plants provide the oxygen necessary for life and sequester the carbon that we produce; microbes and invertebrates cycle the chemicals essential for crop growth; birds and insects pollinate our crops. There are many more examples where biodiversity plays a critical role in human survival, some of them complex. The conclusion is, however, simple: risk to biodiversity can pose a very serious risk to human society.

3   Science and Background

The risks facing biodiversity are complex, individual species tend to face multiple risks and the severity of direct risks are usually determined by one or more driving forces. These drivers are usually socio-economic in nature as are many of the direct risks affecting biodiversity. In order to develop ways of managing risk to biodiversity, we need to consider both the direct risks and their socio-economic drivers. This requires an interdisciplinary approach, involving disciplines such as ecology, toxicology, economics and social science. The topic of risk to biodiversity can best first be illustrated by examples where individual species are at risk from relatively simple pressures. We will consider some examples of species where they are at risk from local human activities, and then consider examples at larger spatial scales where multiple risks affect species and ecosystems. We will also look at the issue of situations where risks are uncertain or ambiguous. In these examples, we will also highlight the issue of conflict in relation to the development of approaches to minimizing risk to biodiversity.

3.1 Simple Risks to Biodiversity

No risk to biodiversity can be considered as "simple." As we saw in the preceding section, species are closely interconnected, therefore a risk to a particular species can have repercussions on a range of other species and the ecosystem processes they perform. For the purposes of the chapter, however, we understand simple risks here as conflicts between particular human activities and individual species.

Birds of prey, or raptors, such as the hen harrier *Circus cyaneus*, the peregrine *Falco peregrinus* or the sparrowhawk *Accipiter nisus* are protected in the UK under the Protection of Wild Birds Act of 1954. Despite this protection, these species are rare and endangered. A major reason for their endangered status is that they are at risk from the actions of farmers and other land managers who consider that these birds of prey in some way threaten their livelihoods. Thus many raptor species are shot, trapped or poisoned (Whitfield et al. 2003).

One notable example is the persecution of birds of prey that consume game birds as part of their diet. There are many such species: Valkama et al. (2005), for example, considered research on the feeding behavior of 52 such predatory species in Europe. In most cases, there is insufficient evidence to support the view that these birds cause significant damage to game birds. The view that birds of prey are a threat to game birds is, however, a common perception of game managers and in some cases research has supported this view.

In the UK, hen harriers and peregrine falcons *Falco peregrinus* has been shown to be able to reduce the size of red grouse *Lagopus lagopus scoticus* populations and the number of birds killed by driven shooting (Redpath et al. 2004). In driven grouse shooting, hunters stand in "blinds" or "butts" while the grouse are driven toward them by lines of human beaters. Grouse densities of over 60 grouse/km^2 are required for this type of hunting, which generates the greatest amount of income for these areas. Large numbers of raptors may lead to significant loss of income for managers of grouse moors and potentially lead to a change in land use to increased densities of sheep and deer or commercial afforestation. This change in land use means a loss of internationally important heather-dominated moorland habitat and the range of upland birds and other species that it contains.

Thus, this is not only an example of a rare species of bird being put at risk because it threatens the livelihoods of certain land managers. Conservation of these birds of prey may also lead to situations where the land managers abandon game bird shooting and change land use to one that has a detrimental impact on biodiversity. Large numbers of raptors therefore present a conservation dilemma (Redpath et al. 2004). This apparently simple example demonstrates the potential complexity of the nature of risk to biodiversity and how conflicts need to be addressed in order to manage this risk.

In direct contrast with birds of prey, the great cormorant *Phalacrocorax carbo* is neither rare nor endangered in Europe, but remains at risk from anglers and commercial fisheries. Indeed, over the last 30 years, great cormorant populations in Europe have expanded both in numbers and geographical range, as a result of increased protection particularly the EC Wild Birds Directive of 1979, the Bern Convention on the Conservation of European Wildlife, the Bonn Convention on the Conservation of migratory species and Wild Animals and the Ramsar Convention on wetlands of international Importance. This bird has also benefited from a non-limiting food supply.

This increase in population size and range, combined with a great ability to fish, has resulted in a number of conflicts between cormorants and commercial fisheries and recreational anglers, angered by the damage caused by cormorants to fishery yields. While some fisheries call for the reduction in size of cormorant populations, conservationists argue that reducing levels of protection could lead to increased risks of illegal killings and unsustainable population levels. An Action Plan for the Management of the Great Cormorant in the African-Eurasian Region has been formulated but has since been largely ignored by individual Member States.

In order to reinforce the Action Plan and reduce the risks to cormorant populations, the REDCAFE project aimed to compile and synthesize all available information on cormorant conflicts and ecology as well as the possible tools for effective conflict management (Carss 2003). This pan-European project provided a novel approach to the conflict as it involved natural and social scientists, as well as all relevant stakeholders to produce an integrated and equitable reporting on the issues. This enabled not only a more complete picture of the conflict issues, but also a better understanding of stakeholder values and opinions.

As the two above examples show, in addition to the many positive effects biodiversity has in terms of use and positive value to humans (as seen in Sect. 2 of this chapter), some biodiversity can be harmful, i.e., have a negative value (Conover 2001). This can take different forms, including damage to livelihoods (e.g., agricultural crops), damage to households, and direct threats to humans. Because of these negative values, certain species face serious threats of persecution.

Large carnivores, for example, have been and still are a controversial subject in many countries in view of their interactions with domesticated animals, game species, and humans (Boitani 2000; Breitenmoser et al. 2000; Vitterso et al. 1998; Vos 2000). In the Indian state of Madhya Pradesh, for example, 1,094 people were injured or killed by wildlife attacks in a 5-year period (Conover 2001). Of these, 121 attacks were from tigers. Because of this direct threat to humans, as well as the poaching of tigers for skins and traditional medicine, and threats from habitat fragmentation, encroachment and developmental projects, the tiger population in India has shrunk to a little over 3,000 individuals. Although the tiger is now considered endangered and protected under the Convention on International Trade in Endangered Species of Wild Fauna and Flora (CITES), tigers continue to be poached. Since 1994, 327 tigers are known to been killed according to the Wildlife Protection Society of India's (WPSI) Wildlife Crime Database. This number, however, is likely to be highly under-representative of the actual extent of tiger poaching.

The reintroduction of predators, such as the white-tailed eagle *Haliaeetus albicilla* in Scotland, for example, have given rise to major conflicts between those in favor of reintroduction and those against it. Some repercussions have included the poisoning, snaring, and shooting of "problem" animals threatening the livelihoods of local farmers. Further studies have indicated that, despite widespread perceptions that white-tailed eagles were directly responsible for high lamb predation in the area, eagle predation was in fact limited at the broad scale, and in many cases targeted on nonviable lambs.

In addition to large carnivores, large herbivores also have important impacts on agriculture. In Africa, for example, a study over a 14-month period reported that elephants raiding crops outside the Amboseli National Park were responsible for causing $200,000 worth of damage to crops and 12 human deaths and injuries (Conover 2001). In India, elephants are responsible for the deaths of 100–200 people a year (Veeramani et al. 1996). Because of these damages, local villagers often see elephants as a serious threat. In a study carried out near Cameroon's Maza National Park, 73% of local villagers thought that more elephants needed to be killed. These negative perceptions pose a very serious threat to the continued survival of elephants across Africa (Tchamba 1996).

In addition to these conflicts between humans and individual species, biodiversity also faces more complex risks that impact on a variety of different species and on different spatial and temporal scales.

3.2   Complex Risks to Biodiversity

Land use change, particularly habitat loss and fragmentation, is the driver expected to have the largest global impact on biodiversity by the year 2100 (Sala et al. 2000). Increasingly, grasslands and forests are being converted into cropland to feed growing food and biofuel demands. Deforestation, in particular, is a major threat. The total area of forest is currently about 4 billion hectares, but it is unevenly distributed with two-thirds of the total forest area found in the ten most forest-rich countries (FAO 2006).

Although its rate is reported to be slowing, deforestation, mainly conversion of forests to agricultural land, is about 13 million hectares per year. The net rate of forest loss is slowed by forest plantations and natural expansion of forests: a net loss of 7.3 million hectares per year is estimated during 2000–2005, compared with a loss of 8.9 million hectares per year from 1990 to 2000. This loss of habitat results in the extinction of local plant species, as well as animal species that depend on plant species composition to survive. The greatest rate of deforestation is in Africa and South America (FAO 2006). Considering that it is estimated that at least 50% of the world's species are found in tropical forests, which cover only 6–7% of the earth's surface (Groombridge 1992), the loss of these forests presents a disproportionate risk to biodiversity.

Even though habitat loss is an obvious threat, the related threat of habitat fragmentation can also have a serious impact to biodiversity. Habitat fragmentation is common throughout the world, although difficult to document. Remote sensing has, however, made it easier to estimate loss and fragmentation of habitats, particularly forest habitats. In Chile, for example, there was an estimated reduction in natural temperate forest area of 67% between 1975 and 2000, and an increase in forest fragmentation, specifically a decrease in forest patch size, a decrease in area of interior forest and a decrease in connectivity among patches (Echeverria et al. 2006).

Some degree of fragmentation is not a serious risk: most species are adapted to patchily distributed habitats or resources. Many species of butterflies, for example, survive in habitat patches within fragmented landscapes. The study of butterflies in these situations, in particular, has given rise to the metapopulation concept (Hanski 1999). In a metapopulation, each local population has a separate probability of extinction and (re)colonization and occupied patches are connected by occasional migration.

Knowledge of the dispersal ability and other characteristics of the ecology of a species together with information on the fragmentation of a habitat can be used to quantify the risk of extinction to particular species in particular landscapes. Schtickzelle et al. (2005), for example, concluded that a patch system occupied by a metapopulation of marsh fritillary butterfly *Euphydryas aurinia* in Belgium could not survive under the present management of the area. Case studies have also shown that fragmentation history is important in determining risk in a changing landscape (Gu et al. 2002) and that the so-called matrix or habitat encountered between patches influences the degree of risk from fragmentation. Grassland butterflies are among those most at risk in this way. Ricketts (2001), for example, in an aptly titled paper "The matrix matters: effective isolation in fragmented landscapes" quantified the resistance to movement of willow thicket and conifer forest to a meadow-inhabiting butterfly community.

In addition to habitat loss and fragmentation, the way that habitats are managed can create threats to biodiversity. Some idea of the multiple threats facing individual species can be seen from the plans that have been put in place to conserve individual species. The UK has developed Biodiversity Action Plans for many species, including the bittern *Botaurus stellaris*, a rare bird species confined to wetlands dominated by the common reed *Phragmites australis*. It became extinct in Britain as a breeding species by 1900 but after recolonization and recovery it started to decline again in the 1950s so that only about 16 pairs were present by the 1990s (Gibbons et al. 1993). Clearly the loss and fragmentation of its habitat are key risks to this species. These were caused by drainage, water abstraction and natural succession to "scrub" habitat due to a decline in reed harvesting. Other threats, however, include water pollution due to agrochemical run-off, pesticide and heavy metal pollution, and salt water intrusion into coastal reedbeds. The main message from this and many other examples is that biodiversity faces multiple risks.

Various risks therefore combine to threaten biodiversity. One example, referred to as "a deadly anthropogenic cocktail" (Travis 2003), is the combination of habitat loss together with climate change, in particular. Indeed, where we would expect that a species is likely to take advantage of climate change by extending its distribution, habitat loss or fragmentation might prevent such an expansion. An analysis of 35 species of butterflies in the UK, all of which were predicted to have expanded their range in response to recent climate change, showed that all but five species had not done so because of a lack of suitable habitat (Hill et al. 2002). Climate change, however complex, is also very uncertain risk to biodiversity, i.e., the short- and long-term effects of climate change are complex and as such not yet fully understood.

3.3   Uncertain Risks to Biodiversity

Climate change is now seen as a major threat to biodiversity, second only to land use change (Thomas et al. 2004; Millennium Ecosystem Assessment 2005). This is partly because of the increasing evidence that the global climate is changing and the growing awareness that climate change has an enormous potential for affecting biodiversity.

Although the Earth's climate has always fluctuated, the current rate of climate change is greater than any experienced during the last 1,000 years, and there is strong evidence that most of this increase is due to anthropogenic climate forcing as a result of increased release of CO_2 (carbon dioxide) and other greenhouse gases into the atmosphere (IPCC 2001; Brooker and Young 2006; Brooker et al. 2007).

The concentration of CO_2 is predicted to rise by 2100 to between 540 and 970 ppm, compared to about 368 ppm in 2000 and about 270 ppm in the pre-industrial era. Climate models predict a rise in temperature of 1.5–5.8°C between 1990 and 2100, which is two to ten times greater than the rise in temperature observed in the last century. Annual precipitation is expected to rise on average by 5–20% over the same period, although the models suggest major regional and seasonal variations. The global mean sea level is expected to rise by 0.09–0.88 m between 1990 and 2100. An increase in the frequency, intensity and duration of extreme events such as more hot days and heavy precipitation events is predicted but the number of cold days is expected to decrease.

The changing climate is already reported to be having an effect on many species. In temperate countries, the bud burst and flowering of plant species is happening earlier, butterfly species are appearing earlier, amphibian and bird species are breeding earlier and migrating bird species are arriving earlier (Walther et al. 2002; Parmesan and Yohe 2003; Root et al. 2003).

There is evidence that birds (Thomas and Lennon 1999) and butterflies (Warren et al. 2001) are extending their ranges polewards. There is evidence of shifts in the range of plants too, although not as rapid (Walther et al. 2005). Although some movement in the distribution of insects and other species is already being detected, it is likely that more significant changes in distribution will occur in the future. Models have been constructed to predict the impact of climate change of the distribution of insects, plants, birds, and other species (Berry et al. 2002). These models demonstrate how some species are likely to contract in range within a country, whereas others are likely to expand their range. Models predicting the impact of climate change on species distribution are based on analysis of species bioclimatic envelopes – the relationship between their observed distributions and climate.

There is some evidence that species can adapt to a changing climate without changing their distribution. Research on four butterfly species within the same family, for example, showed that those living in dry and open habitats had a maximum fecundity and survival rate at a higher temperature than the shade-dwelling species studied (Karlsson and Wiklund 2005). Moreover, populations of the same species have been shown to be adapted to the different habitats they occupy. Populations of the woodland butterfly *Pararge aegeria* living in shady woodland landscape have a higher fecundity at lower temperatures than those originating from open agricultural landscapes and the opposite is true at higher temperatures (Karlsson and Van Dyck 2005).

The response of individual species to climate change should not be considered in isolation: its impact will depend upon interactions between other species of the same and different trophic levels (Bale et al. 2002). Changing temperatures can also have an indirect influence on species that interact with other species that have a different capacity to change their distribution in response to climate change. Although the responses of particular species to changes in climate are poorly known, it is probable that different species will respond in different ways and to different degrees; this will lead to changes in the balance between competing species, plants and herbivores, and insects and their predators (Lawton 1995).

Thomas et al. (2004) used species–area models, an approach discussed in more detail later in the chapter, in relation to predicting the effect of habitat loss on biodiversity, to predict the impact of both climate change and habitat loss on global extinction rates of butterflies and other species. They argued that extinctions arising from reductions in area should apply not only to habitat loss per se but also to climatic unsuitability of that habitat. On this basis, they predicted that for midrange climate scenarios, 15–37% of the taxa they studied would become committed to extinction by 2050. Three different modeling methods were used for three different climate scenarios and they did the analyses twice, once assuming no capacity to disperse and one assuming the species could disperse. This produced a wide range of predicted extinctions. They also predicted that habitat loss alone would result in extinction rates by 2050 of 1–29% in the areas they studied. Despite criticisms of their analysis, including concern about the use of species–area models (Buckley and Roughgarden 2004), Thomas et al. (2004) concluded that climate change represents the greatest threat to biodiversity in most if not all regions of the world.

3.4   Ambiguous Risks to Biodiversity

Many risks are indiscriminate, perhaps affecting only a few species but often affecting many. Certain types of birds, for example, are particularly at risk from nonindigenous invasive species. Often referred to as invasive alien species, these are species that have been deliberately or accidentally introduced outside their native habitats, and have the ability to establish themselves, successfully compete with native species and spread in their new environments. Despite our increasing knowledge of the threat of invasive species, they continue to be a major risk to biodiversity (Hulme 2003). The most infamous invasive species are rats, which have posed a serious risk to island birds and other species poorly adapted to face this risk. Ship rats *Rattus rattus*, Norway rats *R. norvegicus* and Pacific rats *R. exulans* have caused declines or extinctions of land-based birds, burrowing seabirds, flightless invertebrates, and ground-dwelling reptiles (Towns et al. 2006). Globally, ship rats alone are thought to have been responsible for the extinction of about 60 indigenous island species.

Island species have also been at risk from introduced species such as cats (Rodriguez et al. 2006) and hedgehogs (Jones et al. 2005a). Indeed, such are the risks to island species that there is a widespread misconception that extinctions have been restricted to islands (Pimm 2002). Biodiversity is threatened in all parts of the world and invasive species pose a risk to many groups of plants and animals. Invasive species include mammals, reptiles, insects, and plants. More than 2,000 species of nonindigenous plants, for example, are established in the continental USA (Vitousek et al. 1997). Invasive plants include weeds of riparian habitats such as Himalayan balsam *Impatiens glandulifera* and giant hogweed *Heracleum mantegazzianum* (Wadsworth et al. 2000). Invasive insects include the yellow crazy ant *Anoplolepis gracilipes* whose super-colonies have occupied over 30% of the 10,000 ha of rain forest on Christmas Island since the 1990s, and killing of indigenous red crabs have brought about "invasional meltdown" (Abbott 2006). One of the most notorious invasive reptiles is the brown tree snakes *Boiga irregularis*, which caused the extinction or serious decline of most of the 25 resident bird species on the island of Guam (Wiles et al. 2003) and is now a major risk to the biodiversity of Hawaii (Burdick 2005).

Invasive species also pose serious risks to human economies. One example is the zebra mussel *Dreissena polymorpha*, native to the Caspian Sea region of Asia. This invasive species has spread rapidly to all of the Great Lakes since the late 1980s, causing billions of dollars worth in damage by covering the underside of boats and docks and by blocking off pipelines, thereby directly impacting water intake pipes used by cities for their water supply, or hydroelectric companies for power generation. Lafferty and Kuris (1996) estimate the economic impact of the European green crab *Carcinus maenas* to be in the region of $44 million/year due to its impacts on commercial shellfish beds and on large numbers of native oysters and crabs.

Invasive species are, however, very ambiguous, in that the characterization of attributes of successful or unsuccessful invasions are often very general and qualitative. In addition, the actual effects of invasions are often difficult to predict. For example, the common cordgrass *Spartina*, accidentally introduced to the Southern England coast from the East coast of America, have the beneficial effect of stabilizing soft coastal mud of tidal mud-flats through their extensive system of roots and rhizomes. As such, *Spartina anglica* has since been planted to stabilize salt marshes thereby protecting foreshores from erosion (Macdonald et al. 1989). Another example of successful introduction is that of the cactus moth *Cactoblastis cactorum* introduced as a biological control agent against the alien prickly pear cacti *Opuntia* in Australia in the 1920s. Since then, however, scientists have expressed concerns regarding the effects of these moths in Florida, where they now threaten the survival of indigenous *Opuntia* species (Zimmermann et al. 2001).

Other introductions also have very mixed results. One such example is the introductions of fish species into European waters. While the introduction of pikeperch *Stizostedion lucioperca* into many Western European lakes has resulted in major commercial fisheries, its release has also led to a collapse in the cyprinid fisheries. In a similar way, while in economic terms the introduction of rainbow trout *Oncorhynchus mykiss* supports major sport fisheries throughout Europe, rainbow trout has caused considerable loss of native species (Cowx 1997).

4 What Do We Do with What We Know?

4.1 Risk Assessment

4.1.1 Species Assessment

Considering the enormous scale of the problem, we know very little about the impact of the major risks to biodiversity. Even our knowledge of which species are at risk is remarkably poor. The most comprehensive assessments of which species are endangered are done by IUCN. The IUCN Red List, which has been in use for over more than 40 years, classifies species in different categories including critically endangered, endangered, and vulnerable (IUCN 2001; Baillie et al. 2004; Rodriguez et al. 2006). Quantitative criteria are used to classify species in these categories. Species classified as critically endangered, endangered, or vulnerable are frequently described together as "threatened."

The original IUCN Red List classification system operated for nearly 30 years and classified threatened species as extinct, "endangered," "vulnerable" or "rare" ("indeterminate" and "insufficiently known") (Mace and Stuart 1994). It was clearly valuable in producing databases of threatened species, providing a basis for setting conservation priorities and monitoring the success of conservation efforts (Miller et al. 1995). However, this system was frequently criticized for being subjective; classification of species by different authorities varied and did not always correspond with actual risks of extinction (Groombridge 1992). Thus the new IUCN classification system, described above, was adopted in 1994 (IUCN 1994).

The 2004 IUCN Red List named 15,589 species threatened with extinction (Baillie et al. 2004). Amongst some groups of organisms, however, the identification of threatened species is made extremely difficult by our lack of knowledge of the ecology, distribution, and abundance of most species. For example, the 2004 IUCN Red List names 559 insect species but this represents only 0.06% of all described insect species. The equivalent figures for plants and vertebrates were 3 and 9%, respectively. However, the detailed procedures adopted by IUCN include an assessment of amount of data available and species are only evaluated if sufficient data exist. Only 771 insect species were, therefore, evaluated for the 2004 report. Thus 73% of all insects evaluated were classified in the threatened categories. Equivalent figures for plants and vertebrates were 70 and 23%, respectively. Although this suggests that insects are threatened to roughly the same amount as better-known taxa, such is the degree of uncertainty about the status of insects that the true percentage of threatened species should be taken as lying somewhere between 0.06 and 73% (Baillie et al. 2004).

The best overall assessments of the sources of risk to biodiversity come from international initiatives such as the Millennium Ecosystem Assessment (2005). This report concluded that humans have changed had a greater impact on ecosystems in the last 50 years than during any other comparable period. This has been due to the rapidly growing demands for ecosystem services, particularly food, fresh water, timber and fuel, and has resulted in "substantial and largely irreversible loss in the diversity of life on Earth." The Millennium Ecosystem Assessment (MA) acknowledges that the changes made to ecosystems have contributed to substantial improvements in the well-being of humans but they have led to the degradation of many ecosystem services and led to a worsening of poverty for some groups of people. The degradation of many ecosystem services and an increased risk of nonlinear changes, are likely to lead to a decline in the benefits that future generations accrue from ecosystems unless these risks are addressed. The MA identified several examples of nonlinear changes. These include fisheries collapse, the impact of species introductions and extinctions, and regional climate change. The best-known example of fisheries collapse, for example, is the Newfoundland Atlantic cod, which collapsed in 1992 after hundreds of years of exploitation and led to the closure of the Newfoundland fisheries.

4.1.2 General Risk Assessment Models

From a biodiversity perspective, there are several models that address risk. The most widely used models are the "pressure-state-response" and "driver-pressure-state-impact-response" (or DPSIR) models (Fig. 2.3) originally developed by the OECD and used for a range of purposes including biodiversity.

To illustrate how the model works, we can use an example such as industrial production. In this model, industrial production acts as a driving force of environmental change. This driver generates pressures on the environment such as discharges of waste in the air or water. In turn, these pressures will impact the state of

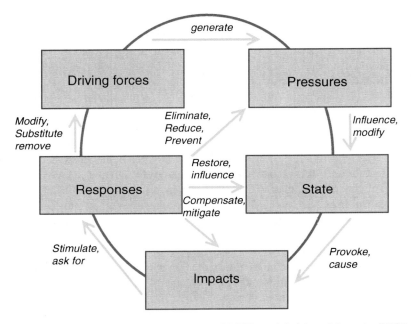

Fig. 2.3 The driver-pressure-state-impact-response (DPSIR) model. Adapted from the OECD

the environment, for example, reducing the water quality of rivers or seas. This change in the state of the environment can cause the water to be unsuitable for drinking, with consequences to the ecosystem, the economy, and the population. The last step in the model is the response to these impacts. In this particular case, a response might be watershed protection, which would then affect all the other steps of the model.

The most interesting aspect of the DPSIR model is that it separates pressures from the drivers that determine the severity of the pressure. Thus, for example, climate change is a direct pressure on biodiversity but the drivers of climate change include various demographic, economic, and socio-political factors. The Millennium Ecosystem Assessment (2005) recognized the same separation but referred to drivers and pressures as indirect and direct drivers of change, respectively.

In addition to the more general models of risks to the environment, there are various approaches to evaluating specific risks to particular species, habitats, or pressures. Below we review a few of these models, specifically addressing risks to biodiversity from habitat loss and fragmentation and invasive species.

4.1.3 Species–Area Models

The risk to biodiversity of habitat loss and fragmentation, for example, is frequently assessed though species–area models. These models are based on the common observation that the number of species inhabiting an area increases as the size of the area increases. This observation was fundamental to the development of "island biogeography" (MacArthur and Wilson 1967) and has been applied to many "island" situations. Indeed, many studies have now shown that the number of species on islands increase as their size increases, whether the study focuses on oceanic islands, soil habitat fragments, or urban roundabouts.

Species–area relationships have been explained through two main hypotheses: the equilibrium theory and the habitat diversity theory. The "equilibrium theory" (MacArthur and Wilson 1967) is that the number of species on an island is a result of the balance between extinction and colonization by new species. Extinction and colonization rates are affected by the number of species on the island, island size, and the distance of the island from other islands and the mainland. Although species are constantly colonizing the island and becoming extinct, an equilibrium is eventually reached which is directly related to island size and inversely related to the distance from other island or mainland sources of species.

The destruction, fragmentation, or modification of habitats poses an obvious risk to biodiversity. Probably the most serious example of this is deforestation, particularly the destruction of tropical forests. As discussed above, what applies to "real" islands often applies to habitat islands. Thus species are likely to become more and more vulnerable to extinction as their habitats become smaller. Species–area relationships may therefore be used to predict the consequences of habitat loss, particularly deforestation, on global species richness (Reid 1992).

The relationship between the number of species (S) in an area and its size (A) can be represented by the equation of a straight line:

$$\log S = \log c + z \log A$$

where c and z are constants.

Studies on islands, mainland areas and habitat islands have shown that the slope (z) generally lies in the range 0.10–0.50 (Lomolino 2000). Reid and Miller (1989) based their predictions of species extinctions on a slope of between 0.15 and 0.40. These slopes predict that a 90% reduction in habitat size will result in the loss of 30–60% of the species present. Reid and Miller assumed a deforestation rate of 0.5–1% loss of forest area per year, one to two times the estimate for 1980–1985 (FAO 1988). The predicted rate of extinction based on these predictions was a loss of 2–5% of species per decade. Most extinction predictions based on habitat loss range from 2–6% to 8–11% species per decade (Mawdsley and Stork 1995). In Singapore, habitat loss exceeded 95% from 1819 to 2002 (Brook et al. 2003). Documented extinctions and inferred extinction rates for butterflies, freshwater fish, birds, and mammals ranged from 34 to 87%. A species–area model predicted that the current rate of habitat loss in South-East Asia would lead to a loss of 13–42% species, of which about a half would be global extinctions.

4.1.4   Invasive Species Models

In view of the potential negative impacts of invasive species on local biodiversity as well as more far-reaching socio-economic aspects, it is essential to develop tools to assess the risks of invasives. One such tool has been developed by the ALARM

Fig. 2.4 Invasives risk
assessment tool. Adapted
from the ALARM project
(http://www.alarmproject.
net/alarm/)

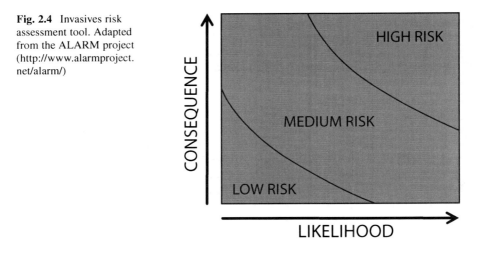

project, which assesses risk of invasives according to the potential consequences of invasive species (economic, social, and environmental impacts as well as management costs) and the likelihood of species becoming invasive (strength of pathways, establishment, population growth, dispersal) based on species distribution data (Fig. 2.4).

4.1.5 Population Viability Analysis

A common approach to evaluating acceptable risk to biodiversity is population viability analysis. Population viability analysis is an ecological assessment: acceptable risk can also be considered from social, legal, and economic perspectives. From social and legal perspectives, acceptable risk relates to individual and societal values toward the conservation of particular species and habitats. For most species such values are irrelevant: most species are either so common or their status is so poorly understood that they are not perceived to be at significant risk. They are not afforded explicit legal protection and few people express concern about risks to their existence despite the fact that most major risks to biodiversity are likely to affect those species that we know very little about. There is, however, increasing public awareness of some relatively common species such as farmland birds. Indeed, the trend in abundance of farmland birds has become the single most important indicator of biodiversity in Europe. It is, of course, a relatively easy measure of biodiversity and it is assumed to be a good indicator of trends in species that are less easy to monitor. Nevertheless, its success as an indicator probably demonstrates the strength of the esthetic and cultural values that many people hold for species that are not endangered.

From an ecological perspective acceptable risk relates to long-term population viability. If the species is abundant and has a high rate of reproduction, then its susceptibility to risk is inevitably less important than if the species is rare. From a social point of view, however, the esthetic and cultural value attached to a particular species has an influence on what may be considered to be acceptable risk. Thus risks to common garden birds may be considered to be less acceptable than risks to rare snakes even though the latter is more susceptible to extinction. The latter example, of course, introduces the direct risk that some species pose to humans, an issue that we will return to later.

Whatever model is used to address risk to biodiversity, it is important to realize that there are factors that have a direct influence on biodiversity and factors that have an indirect influence by determining the magnitude of the direct influences. All of these indirect influences (or drivers) are associated with human action as are many of the direct influences. Thus, for example, the predatory birds that are a major example in *Silent Spring* continue to be at direct risk from pesticides residues in their food. However, the pesticides are applied to provide an economic return from agriculture, which is itself driven by human population size and the demands of that population for a specific amount and quality of food. The application of pesticides is not always economically effective, excessive food production is common in developed countries and food quality is often set by appearance rather than nutritional quality. Nevertheless, the main driving force behind pesticide application is human demand for food. Indeed, the apparently irrational behavior just referred to, which leads to more pesticide application than would appear to be necessary, is driven by factors such as economic risk averse attitudes in farmers, national policies that encourage excessive food production and public attitudes to food appearance. Thus the direct risk of pesticides to predatory birds is driven by a complex set of driving forces, each linked to demands and attitudes of human society.

4.2 Risk Management

4.2.1 Biodiversity Policy

Public concern for biodiversity has risen in the last 50 years and has led to national and international policies, legislation, and actions to conserve biodiversity, notably the Convention on Biological Diversity (CBD). For an overview of major landmarks in international biodiversity policy, see Table 2.1.

In Europe, the first Environmental Action Programme was launched in 1973, the Birds Directive in 1979 and the Habitats Directive in 1992. However, the CBD led to a rapid increase in the development of policy on biodiversity in Europe and elsewhere (see Table 2.2 for a full overview of the major landmarks in EU biodiversity policy since 1979).

Table 2.1 Major international biodiversity policy landmarks

Date	Instrument	Aims	Status
1971	The Convention on Wetlands of International Importance especially as Waterfowl Habitat ("Ramsar Convention")	Wetland conservation and wise use	Came into force 1975, 152 Parties (as of 1 July 2006)
1973	Convention on International Trade in Endangered Species of Wild Fauna and Flora (CITES)	Control of international trade in specimens of wild animals and plants	Came into force 1975, 169 Parties (as of 1 July 2006)
1979	The Convention on the Conservation of Migratory Species of Wild Animals ("Bonn Convention")	Conservation of terrestrial, marine and avian migratory species throughout their range	Came into force 1983, 97 Parties (as of 1 May 2006)
1979	Convention on the Conservation of European Wildlife and Natural Habitats ("Bern Convention")	Conservation of wild flora and fauna and their natural habitats	Came into force 1982, 45 Parties (as of 1 March 2005)
1992	Convention on Biological Diversity (CBD)	Biodiversity conservation; sustainable use of biodiversity and equitable benefit sharing	Came into force 1993, 188 Parties (as of 1 July 2006)

At the national scale, biodiversity strategies have been published in many countries including Sweden (1994), UK (1994), Slovenia (2001), and Slovakia (1997) and are in preparation elsewhere. The European Community published a biodiversity strategy in 1998, based on a policy of incorporating biodiversity concerns in sectoral policies and adopted four Biodiversity Action Plans in 2001 – on the Conservation of Natural Resources, Agriculture and Fisheries and on Economic and Development Co-operation.

However, realizing that current policies and action taken to conserve biodiversity were inadequate, the European Union at its 2001 summit meeting in Göteborg, Sweden, set the ambitious target to "protect and restore habitats and natural systems and halt the loss of biodiversity by 2010." A similar target was set by the CBD in 2002 "to achieve by 2010 a significant reduction of the current rate of biodiversity loss" and endorsed by the World Summit on Sustainable Development (WSSD) in 2002. In response, the European Commission published a detailed policy communication on biodiversity in 2006. In many cases, especially where habitats or species are particularly vulnerable to risk, protected areas and species protection are adopted as viable risk management methods.

Table 2.2 Major landmarks in EU biodiversity policy

Date	Instrument	Remarks
1979	Directive on the Conservation of Wild Birds (79/409/EEC)	The Directive requires Member States to identify and manage areas of conservation for birds
1992	Directive on the conservation of natural and seminatural habitats and of wild flora and fauna (92/43/EEC)	The Directive requires Member States to identify and manage areas of conservation for selected species and habitats
1992	Agri-environment Regulation 2078/92	Requires Member States to apply agri-environment measures where appropriate
1998	"Cardiff" process of environmental integration	Strategy setting out guidelines to integrate the environmental dimension into other policies. Nine sectoral strategies are presented (agriculture, transport, energy, industry, internal market, development, fisheries, economics and finance and foreign affairs)
1998	Sustainable Development Strategy	The strategy sets objectives, targets, and concrete actions for seven key priority challenges for the coming period until 2010, including the better management of natural resources
1998	European Community Biodiversity Strategy	The strategy defines the framework for defining Community policies and instruments to comply with the CBD
2001	Biodiversity Action Plans in the areas of Conservation of Natural Resources, Agriculture, Fisheries, and Development and Economic Cooperation	Four Action Plans define concrete actions and measures to meet the objectives defined in the European Community Biodiversity Strategy, and specify measurable targets
2006	EC Communication on "Halting the loss of biodiversity by 2010 and beyond"	The Communication sets out 10 policy objectives in 4 policy areas: Biodiversity in the EU; The EU and global biodiversity; Biodiversity and climate change; and The knowledge base

4.2.2   Protected Areas

Potentially feasible management options exist for some risks to biodiversity, other risks are much more difficult to manage. Protected areas have been established mainly to minimize the risk of habitat loss to biodiversity. There are more than 104,000 protected areas in the world, covering over 12% of the land area, but only about 0.5 and 1.4% of the ocean and the marine coastal zone, respectively (Chape et al. 2005). The modern era of protected areas started when the Yellowstone National Park was established in 1872. A protected area has been defined by the IUCN and the World Commission on Protected Areas (WPCA) as "an area of land and/or sea especially dedicated to the protection and maintenance of biological diversity, and of natural and associated cultural resources, and managed through legal or other effective means." This definition covers many different types of protected areas; in response to these differences the IUCN proposed six categories including those afforded the strictest protection (Category I). The Yellowstone National Park, established in 1872, for example, is an IUCN Category II protected area, a national park. Protected area programs include the UNESCO World Heritage Sites, UNESCO Man and the Biosphere (MAB) sites, ASEAN Heritage Parks and Reserves, sites established under the Ramsar Convention on Wetlands, and the European Natura 2000 network. Despite the number of protected areas some ecosystems remain poorly protected, particularly those experiencing the greatest risk (Hoekstra et al. 2005); only about 5% of the tropical humid forests, for example, are protected to some degree (UNEP-WCMC 2003).

Without the threat of climate change and the decline in the degree to which species can move readily across the landscape between protected areas, protected areas may be considered to be an adequate management option. However, protected areas are unlikely to be able to protect biodiversity from all the risks it faces. And, as discussed below, the establishment of protected areas is often done in such a way as to alienate the local population and increase the risk to biodiversity.

Despite the many advantages of protected areas many rare species exist outside them and their continued survival requires the presence of sustainable populations in nonprotected areas. In Europe, the Birds and Habitats Directives aim to address this issue by creating Natura 2000, a pan-European network of Special Areas of Conservation and Special Protection Areas to ensure the favorable conservation status of habitats and species in their natural range. The network of Natura 2000 sites is illustrated in Fig. 2.5.

Natura 2000 is, however, not a network of strictly protected areas but rather a network of areas in which active steps are being taken to reconcile biodiversity conservation in Europe with the need to "take account of economic, social and cultural requirements and regional and local characteristics" (Article 2(3) of the Habitats Directive). This is particularly important in the context of Natura 2000, where the sites selected are in the main owned and, more importantly, managed by private landowners. However, despite this drive to include local actors in the development of Natura 2000, this initiative has seen a number of conflicts, leading to serious delays in its implementation. Participants of the Bath conference on Natura 2000 and People 1998 identified the "resistance of local people concerned that their economical and social interests might be threatened by the designation of a site" as one of the main reasons for the delay in implementing the Natura 2000 network. One extreme example is the "Groupe de 9" in France, who questioned the legitimacy of the implementation in France and ultimately caused the national suspension of the directive in 1996 (Alphandery and Fortier 2001). Other examples of conflicts caused by the Directive include Finland (Sairinen et al. 1999) and Germany (Wagner 2000). Preliminary lessons learned from these delays and conflicts are that increased communication and transparency at every stage of implementation with local actors is essential for the acceptance and future involvement of local communities in biodiversity conservation.

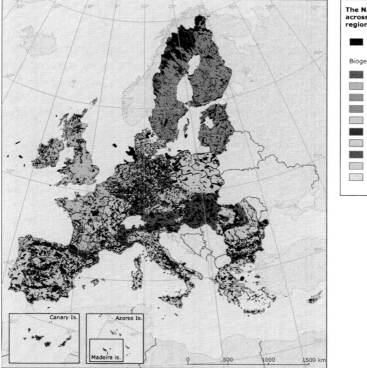

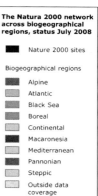

Fig. 2.5 Natura 2000 coverage across biogeographical regions. *Source*: European Environment Agency

4.2.3   Species Protection

After protected areas, the most common option for managing risk to biodiversity is species protection. Many species are afforded legal protection, either directly or indirectly through legal restrictions on trade. The latter includes CITES, the Convention on International Trade in Endangered Species of Wild Fauna and Flora.

Birds of prey are protected by national and international legislation. But, as with many species and habitats, legislation is only partially effective: either the risks are too complex to be suitably addressed by legislation or it is too difficult to adequately enforce the legislation. Legislation will also usually fail to address the driving forces that created the risk that the species faces. Thus, laws protecting birds of prey usually do not address the economic damage that they do or are perceived to do to livestock or gamebirds.

Consequently, a further management option is to use methods that minimize the economic damage done by birds of prey and other threatened species. Research on the hen harrier has showed that supplementary feeding may be useful in reducing the number of grouse chicks taken by these birds of prey (Redpath et al. 2001). It is unlikely that methods such as supplementary feeding could be developed for many threatened species. In some cases, threats to rare species can be managed by compensating farmers or other land managers for the actual or potential income lost by the activities of those species. According to a cost–benefit analysis on overwintering brent geese *Branta bernicla* grazing on farmers' crops in Britain, compensation was found to be the optimal financial solution for farmers (Vickery et al. 1994). This has been put in place in some countries, such as Estonia, where farmers can be compensated for losses incurred due to migrating birds feeding on cereal fields.

Compensations paid for attacks by carnivores on livestock vary greatly according to country, ranging from no compensation at all, to direct compensation for damages incurred, to subsidies paid for implementing measures to reduce risks of attacks. Another possibility is to pay farmers according to the risks within the area they live. This could hypothetically mean that provided that a farmer is rigorous about protecting his livestock against large carnivore attacks he could actually gain financially from this system.

Legal protection of species and their habitats requires knowledge of which species are most at risk, a process led internationally by IUCN and nationally by many government agencies, NGOs, and researchers. Monitoring plays a key role in identifying which species and habitats are most at risk and in estimating trends in the severity of particular risks. Biodiversity monitoring activities have increased in recent decades. The number of species monitored in detail is, however, limited to better-known groups of species. For example, the identification of endangered species has been successful for well-known groups of species such as birds (Lei et al. 2003) and butterflies (Wenzel et al. 2006), and the process itself has clearly also brought more effective nonregulatory action (De Juana 2004). Nevertheless, monitoring of some species has been important in exposing the emergence of new risks.

Monitoring can also be focussed on particular risks such as DDT and other persistent organochlorine pesticides. Although *Silent Spring* led to their restricted use in many countries, these chemicals are still widely used and are detected in monitoring of bird tissue (e.g., Kunisue et al. 2002). Monitoring is also done to assess exposure to chemicals such as polychlorinated biphenyls (PCBs) (Wienburg and Shore 2004). However, many groups of species are too poorly known for accurate assessments of risk to be possible.

Regulation will continue to play an important role in managing the risks that threaten biodiversity. The legal protection of species and their habitats, and legal restrictions on activities that put them at risk, are all necessary, although unlikely to be completely effective (e.g., Holdich and Pockl 2005).

Regulatory actions that focus on particular species are most suitable where the species is directly threatened by human action but, as discussed above, regulation is unlikely to be completely effective where human livelihoods are themselves threatened. Regulatory actions that focus on the risk rather than on species and habitats have the considerable advantage of potentially providing protection for many species. Many of the major risks to biodiversity, however, do not easily lend themselves to regulation, notably climate change. The risks posed by invasive species have, however, been frequently addressed by regulation (Stohlgren and Schnase 2006).

4.2.4 Conflict Management

The successful management of risk to many threatened species and habitats requires an acknowledgment that human activities are a significant source of that risk and that, therefore, the conservation of these species and habitats is in conflict with these activities. Thus, for example, the conservation of birds of prey is often in conflict with activities such as livestock farming, hunting, fishing, pigeon racing, and wind farming.

In an attempt to explore ways of managing the conflict between the conservation of a legally protected raptor, hen harrier *Circus cyaneus*, and the commercial hunting of red grouse *Lagopus lagopus scoticus*, Redpath et al. (2004) quantified the views of stakeholders on raptors and management options using hierarchical decision trees and multicriteria decision analysis. The decision-modeling exercise not only reengaged stakeholders in a dialog, but also allowed stakeholders to better understand different, and potentially conflicting, perceptions of raptors.

As the examples discussed above demonstrate, the management of risks to biodiversity cannot therefore be adequately done without consideration of the impact of biodiversity on stakeholder livelihoods, the perceptions of these stakeholders and their actions. Stakeholder perceptions of the issue and of other stakeholders involved in the process are essential aspects to consider in all risk management options. The perceptions of groups of stakeholders can be very strong and can negatively affect communication among groups, sometimes so much so that certain groups will face disapproval before any direct contact has even taken place (Stoll-Kleemann and O'Riordan 2002).

The main drivers of conflict in Europe are related to changes in land use. Such changes include the intensification or abandonment of silvicultural and agricultural practices, recreation and hunting, and policy-related threats, particularly policies such as the Common Agricultural Policy and EU environmental Directives including the Wild Birds Directive and Habitats Directive (and the associated Natura 2000 network mentioned above). Conflicts in themselves vary tremendously according to the issues and stakeholders involved in the conflict. However, even though the choice and applicability of conflict management strategies will depend on the dimension of a conflict, a number of potential biodiversity conflict management strategies exist (Jones et al. 2005b; Young et al. 2005).

Biodiversity conflict management options can include political, economic, or legislative means to reduce biodiversity conflicts, ranging from the provision of incentives for biodiversity conservation such as agri-environmental schemes under the CAP, to EU level initiatives aiming to legislate for the conservation of biodiversity such as the Natura 2000 network. As with most conflict management strategies, political, economic, or legislative means can, however, lead to further exacerbation of conflicts, with stakeholders potentially resisting such top-down approaches unless they feel they have a level of control over the creation and implementation of laws or regulations. Other potential conflict management options are the buying or leasing of land for conservation purposes and the application of strict Environmental Impact Assessments (EIAs) to identify and avert negative effects of infrastructure projects and thus potentially defuse conflicts. The application of new technologies, management practices, or land-use patterns, applying spatial planning methods (Nowicki et al. 2005) and other techniques can all contribute to conflict management processes.

Deliberative and inclusionary processes such as community-based management, communication and dialog, educational programs and co-management planning can assist conflict management through the inclusion of different stakeholders, sharing of common visions and positive social capital that is built by different parties working together on practical projects.

4.2.5 Increased Research

As argued above, the management of risk to biodiversity requires attention to the direct drivers of biodiversity loss such as chemical pollution and species persecution and their underlying causes, which are almost always human in origin. Risk management therefore often means conflict management. However, the sustainable management of conflict and the successful reversal of biodiversity loss requires that their underlying causes are understood. This needs a much better understanding of public attitudes to biodiversity and its loss.

However, although public involvement in biodiversity management is now heralded as essential for the success of conservation initiatives and the reduction in conflicts between conservation and other human activities, scientists and policy makers often argue that public knowledge of biodiversity and biodiversity conservation options is insufficient to allow for effective decision-making. A number of studies have refuted this, including a recent study in the Cairngorms area of Scotland that found that a cross-section of members of the general public expressed rich views and concepts of biodiversity, which in turn informed their attitudes on biodiversity conservation (Fischer and Young 2007). Although the participants of the study had, in many cases, a little knowledge of the term biodiversity, they developed very complex constructs of biodiversity informed by the normative and idealistic components of biodiversity concepts.

Despite increased interest in public attitudes and public participation in environmental decision-making, more research is required on interpreting biodiversity concepts in order to better understand public attitudes to biodiversity and better involve the public in conservation options. This increased involvement could be instrumental in bridging the gap between conflict and the sustainable use of biodiversity.

4.3 Risk Communication

The communication of risks to biodiversity to the public is not a new phenomenon, and has gained ground in certain years, particularly with issues such as climate change increasingly making it to the front page of newspapers. However, these reports are often presented in the media in a rather dramatic way, leading to confusing and often seemingly conflicting messages.

Table 2.3 Two models of communication

"Marketing" model	"Education" model
Treats environmental information as a "brand"	Treats environmental information as an aid to understanding
Focuses on presenting messages in an attractive format, rather than on content	Focuses on providing accurate content
Success is measured in terms of "brand awareness"	Success is measured in terms of understanding
Eliminates contextual information in order to present a simple message	Ensures that contextual information is provided in order to facilitate interpretation

Source: Adapted from C. Richards, RISK Workshop, Banchory 30th October 2007

Perhaps because of these problems and other pitfalls of communication, research scientists rarely engage in the communication of their research to the wider public, leaving others to communicate the implications of their research for biodiversity on their behalf. Communication of environmental research to the public therefore often implicitly or explicitly adopts a "marketing model," aiming to convey simple messages "as a brand that can be sold," as advocated by the Institute for Public Policy Research (Ereaut and Segnit 2006). Approaches such as those more typically used in education, which aim to develop the capacities of audiences to evaluate competing messages and to understand the reasoning behind them, are employed much less often in the communication of research findings, but seem likely to be of more lasting value. This has led to the development of a number of models of communication, including, among other the marketing model and the education model (see Table 2.3).

These various models all aim to communicate risks. However the ways in which these are communicated vary greatly from alarmism to lists of particular actions needed to alleviate risks, and processes to develop capacities and understanding. The result in the case of a complex situation like the communication of climate change is a "very messy and noisy language landscape with advocates apparently arguing among themselves in the battle for consensus" (Ereaut and Segnit 2006).

In summary, current approaches to biodiversity risk communication still lack a clear understanding as to the steps required in developing a framework to create a successful communication strategy where risks to biodiversity are presented in such a way that the information (including uncertainties) is clearly communicated to all relevant groups of people. From the above, it would appear that three main steps to achieving this communication mechanism are necessary, including the identification of current barriers to communication; the recognition of risks of communication and finally; and the evaluation of the effectiveness of communication.

4.3.1 Barriers to Communication

The majority of issues relating to biodiversity and risks to biodiversity are remarkably complex not only in themselves, due to the relationships between components of these systems, but also due to the multitude of interlinkages among environmental phenomena (van den Hove 2000). This complexity is often reflected in the language used by research scientists and is often difficult to translate into simpler terms.

In addition, a certain amount of uncertainty exists in all scientific research including extrinsic uncertainties (i.e., insufficient scientific knowledge) and intrinsic uncertainties inherent to the complexity and indeterminacy of environmental issues (van den Hove 2000). When communicating research results, scientists often feel compelled in the light of ethical considerations to highlight these uncertainties, even in cases where the uncertainty is comparatively low. However, once communicated, these voiced uncertainties can potentially lessen the impact of particular messages.

4.3.2 Risks of Communication

As the example above shows, a major risk associated with current communication of risks to biodiversity can lead to simplistic or hysterical messages to the public, which can in turn become saturated and jaded. In some cases, scientists also choose to present their finding in a more sensationalist manner. However, are they then presenting risk in an unethical way? Indeed, what are the rights and responsibilities of scientists in communicating risks? One major consideration is the examination of the reasons behind communication. Indeed, the current assessment of scientists in many universities and research institutes relies on the number and quality of publications in peer-reviewed journals. These are very rarely read in their current forms by the wider public. In this respect, not only are the incentives in place for scientists to communicate to the wider public lacking, but the abilities and skills to communicate effectively are also absent.

4.3.3 Evaluating the Effectiveness of Communication

As highlighted above, communication of risks to biodiversity is currently taking place, with, in some cases, very high budgets spent on communicating to certain sectors of the population. Despite these efforts, very little evaluation of the effectiveness of communication has or is taking place. Although some studies have recently highlighted the ability of the public to understand complex biodiversity issues including risks to biodiversity, little is known on the willingness of these different groups to understand these issues. In addition, little is known on the impact of different information sources on public perceptions and beliefs and indeed on how these different information sources vary from the source to the recipients of such information.

5 Future Perspectives

Risk to biodiversity is largely associated with global change, the nature and degree of which has been impossible to predict over the last 50 years. Although we know an increasing amount about climate change, land use change, the spread of invasive species, and other factors that pose risk to biodiversity, we cannot accurately predict their future impact. Increasingly, however, plausible alternative scenarios are being produced, coupling possible environmental trends with possible responses.

Sala et al. (2000) developed three alternative scenarios of change for the year 2100 based on scenarios of change in atmospheric carbon dioxide, climate, vegetation, and land use and known sensitivities of biodiversity to these parameters. These scenarios were based on the assumption of no interactions among causes of biodiversity change, antagonistic interactions and synergistic interactions. In all scenarios, grasslands, and Mediterranean ecosystems experience the greatest loss of biodiversity due to their sensitivity to all drivers of change. In the first scenario, based on the assumption of no interactions among causes of biodiversity change, changes among biomes are relatively small. In the second scenario, based on the assumption of antagonistic interactions, tropical and southern temperate forests, and arctic ecosystems experience the greatest changes due to land use change (for forests) and climate change (in the arctic). In the last scenario, based on the assumption of synergistic interactions, grasslands and Mediterranean ecosystems are most affected, with the range of expected changes in other biomes expected to be broader than in the first scenario. This kind of study is increasingly important to understand drivers of change and their potential impact on biodiversity. However, the authors warn that scenarios such as these need to be further refined through quantitative regional analysis and research on the interactions between drivers in order to be useful for policy makers.

The Millennium Ecosystem Assessment (2005) has also produced four contrasting scenarios (Box 2.1). Each of their scenarios has a different effect on biodiversity, as shown below in Fig. 2.6.

Box 2.1 Millennium Ecosystem Assessment scenarios

Global Orchestration – This scenario depicts a globally connected society that focuses on global trade and economic liberalization and takes a reactive approach to ecosystem problems but that also takes strong steps to reduce poverty and inequality and to invest in public goods such as infrastructure and education. Economic growth in this scenario is the highest of the four scenarios, while it is assumed to have the lowest population in 2050.

Order from Strength – This scenario represents a regionalized and fragmented world, concerned with security and protection, emphasizing primarily regional markets, paying little attention to public goods, and taking a reactive approach to ecosystem problems. Economic growth rates are the lowest of the scenarios (particularly low in developing countries) and decrease with time, while population growth is the highest.

Adapting Mosaic – In this scenario, regional watershed-scale ecosystems are the focus of political and economic activity. Local institutions are strengthened and local ecosystem management strategies are common; societies develop a

---

(continued)

Box 2.1 (continued)

strongly proactive approach to the management of ecosystems. Economic growth rates are somewhat low initially but increase with time, and population in 2050 is nearly as high as in Order from Strength.

TechnoGarden – This scenario depicts a globally connected world relying strongly on environmentally sound technology, using highly managed, often engineered, ecosystems to deliver ecosystem services, and taking a proactive approach to the management of ecosystems in an effort to avoid problems. Economic growth is relatively high and accelerates, while population in 2050 is in the midrange of the scenarios.

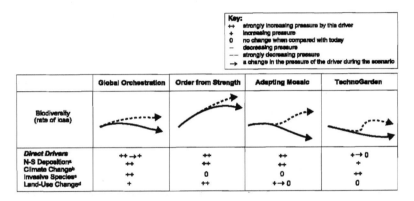

Fig. 2.6 Rate of loss of biodiversity in four contrasting scenarios. The *solid line* indicates the best case and the *dashed line* indicates the worst case envisioned for each scenario. *Source*: Millennium Ecosystem Assessment (2005)

6 Chapter Summary

Risk to biodiversity is inevitable but the natural resilience of species and habitats means that adaptation to these risks is continually occurring. For this reason and because of the poor knowledge of the many species that comprise biodiversity, the realization of the scale of its loss is relatively recent. Action to reduce these risks have, however, been going on for a long time, notably through protected areas, legislation to protect particularly threatened species and habitats, and measures to minimize the risk of particular threats. The management of risk to biodiversity has therefore been very responsive and has had many successes. But it has often lacked a strategic element so that, it may be argued, some of the most threatened species and the most serious risks have not been adequately addressed.

This situation has changed due, in particular, to the Convention on Biological Diversity, the actions of NGOs including IUCN, the emergence of a new approach to protected areas, the Millennium Ecosystem Assessment and the development of a more strategic approach to research on biodiversity.

A more strategic approach to the management of risk to biodiversity is now possible. Although this would have had a much more effective if implemented 30 years ago, it was inevitable that the awareness of biodiversity loss and its causes would be a gradual process. Moreover, there is much more to be learnt about trends in biodiversity and the risks it faces.

Although better monitoring of biodiversity, better assessment of risk and a more strategic approach to conserving biodiversity are all essential elements to successful risk management, an equally pressing need is the effective and targeted communication of risk to the public, policy makers, and other stakeholders. As demonstrated in this chapter, risk to biodiversity and to humans is complex and constantly changing. As such, future perspectives developed by scientists from a range of disciplines including natural sciences and social sciences need to be continually developed, updated, and communicated effectively in a timely manner to policy makers and the public.

Glossary

Acid rain: The deposition of sulfur compounds in rain and other forms of precipitation, after sulfur dioxide emitted into the atmosphere undergoes chemical transformation and is absorbed in clouds.

Alien species: In a given ecosystem, an alien species is any species that is not native to that ecosystem.

Biodiversity: The variability among living organisms from all sources including, inter alia, terrestrial, marine, and other aquatic ecosystems and the ecological complexes of which they are part; this includes diversity within species, between species and of ecosystems.

CO_2 (carbon dioxide): A gas produced during the combustion of carbon-based compounds such as fossil fuels and wood (and during respiration). CO_2 contributes to climate change as a greenhouse gas.

Convention on Biological Diversity (CBD): The first global agreement on the conservation and sustainable use of biological diversity, signed by world leaders at the United Nations Conference on Environment and Development in Rio de Janeiro, Brazil in 1992.

Ecosystem: A dynamic complex of plants, animals, and micro-organisms and their nonliving environment interacting as a functional unit.

Ecosystem approach: A strategy for the integrated management of land, water, and living resources that promotes conservation and sustainable use in an equitable way.

Ecosystem functioning: The collective activities of plants, animals, and microbes and the effect these activities have on the physical and chemical conditions of their environment.

Ecosystem goods: Physical elements that are directly or indirectly consumed by humans.

Ecosystem services: Services provided by the movement of energy and nutrients through the air, water, and land and through the food chains.

Endangered: A species is endangered when it is facing the risk of becoming extinct.

Endemic: Unique to or confined to a particular area.

Evapotranspiration: The process of transferring moisture from the earth to the atmosphere by evaporation of water and transpiration from plants.

Extinct: A species is extinct when there is no reasonable doubt that the last individual of that species has died.

Fragmentation (of habitat): The conversion of large areas of contiguous habitat such as native forest to other types of habitat leaving remnant patches of habitat (e.g., forest) that are isolated from each other. Habitats may also become fragmented by roads and other forms of transport. This may have a detrimental effect on species requiring a certain minimum area and species may also face greater risks when dispersing across fragmented, rather than un-fragmented, landscapes.

Invasive species: A species, usually an alien species, whose introduction does or is likely to cause economic or environmental harm or harm to human health.

Monitoring: The continuous investigation of a given population or subpopulation, and its environment, to detect changes in numbers, distribution, or behavior.

Nitrogen deposition: The input of nitrogen to terrestrial and aquatic habitats from the atmosphere (in precipitation, aerosols or in gaseous forms).

Rate of extinction: The background rate of extinction is the number of extinctions that would be occurring naturally in the absence of human influence. The current rate of extinction is the number of extinctions that are currently occurring in the presence of human influence.

Sequestration of carbon: The process of removing carbon dioxide from the atmosphere.

Species: Group of organisms that have a high degree of physical and genetic similarity, and that generally interbreed only among themselves.

Sustainable use: The use of components of biological diversity in a way and at a rate that does not lead to the long-term decline of biological diversity, thereby maintaining its potential to meet the needs and aspirations of present and future generations.

References

Abbott KL (2006) Spatial dynamics of supercolonies of the invasive yellow crazy ant. *Anoplolepis gracilipes*, on Christmas Island, Indian Ocean. *Diversity and Distributions* 12(1):101–110

Alphandery P, Fortier A (2001) Can a territorial policy be based on science alone? The system for creating the Natura 2000 network in France. *Sociologia Ruralis* 41:311–328

Baillie EM, Hilton-Taylor C, Stuart SN (eds) (2004) *2004 IUCN Red List of threatened species. A Global Species Assessment.* IUCN, Gland, Switzerland and Cambridge, UK

Bale JS, Masters GJ, Hodkinson ID, Awmack C, Bezemer TM, Brown VK, Butterfield J, Buse A, Coulson JC, Farrar J, Good JEG, Harrington R, Hartley S, Jones TH, Lindroth RL, Press MC, Symrnioudis I, Watt AD, Whittaker JB (2002) Herbivory in global climate change research: direct effects of rising temperature on insect herbivores. *Global Change Biology* 8(1):1–16

Berry PM, Dawson TP, Harrison PA, Pearson RG (2002) Modelling potential impacts of climate change on the bioclimatic envelope of species in Britain and Ireland. *Global Ecology and Biogeography* 11(6):453–462

Boitani L (2000) *Action Plan for the conservation of wolves (Canis lupus) in Europe.* Council of Europe Publishing, Strasbourg

Bowerman WW, Best DA, Grubb TG, Zimmerman GM, Giesy JP (1998) Trends of contaminants and effects in bald eagles of the Great Lakes basin. *Environmental Monitoring and Assessment* 53(1):197–212

Breitenmoser U, Breitenmoser-Würsten C, Okarma H, Kaphegyi T, Kaphegyi-Wallman U, Müller UM (2000) *Action Plan for the Conservation in Europe of the Eurasian Lynx.* Council of Europe Publishing, Strasbourg

Brook BW, Sodhi NS, Ng PKL (2003) Catastrophic extinctions follow deforestation in Singapore. *Nature* 424:420–423

Brooker R, Young J (eds) (2006) *Climate change and biodiversity in Europe: a review of impacts, policy responses, gaps in knowledge and barriers to the exchange of information between scientists and policy makers.* Final Report for Department for Environment, Food and Rural Affairs, UK. http://www.defra.gov.uk/wildlife-countryside/resprog/findings/climatechange-biodiversity/report.pdf. Accessed 12 Feb 2007

Brooker R, Young J, Watt A (2007) Climate change and biodiversity: impacts and policy development challenges – A European case study. *International Journal of Biodiversity Science and Management* 3(1):12–30

Buckley LB, Roughgarden J (2004) Biodiversity conservation – Effects of changes in climate and land use. *Nature* 430(6995):1–119. doi:10.1038/nature02717

Burdick A (2005) *Out of Eden: An Odyssey of Ecological Invasion.* Farrar, Straus and Giroux

Carss DN (ed) (2003) Reducing the conflict between Cormorants and fisheries on a pan-European scale - REDCAFE Final Report. Natural Environment Research Council, Centre for Ecology and Hydrology, Banchory, Scotland, UK

Chape S, Harrison J, Spalding M, Lysenko I (2005) Measuring the extent and effectiveness of protected areas as an indicator for meeting global biodiversity targets. *Phil Trans R Soc Lond B* 360:443–455

Conover MR (2001) Resolving human-wildlife conflicts: The science of wildlife damage management. Lewis Publishers, Boca Raton, Florida

Costanza R, d'Arge R, de Groot R, Farber S, Grasso M, Hannon B, Limburg K, Naeem S, O'Neill RV, Paruelo J, Raskin RG, Sutton P, van den Belt M (1997) The value of the world's ecosystem services and natural capital. *Nature* 387:253–260

Cowx IG (1997) Introduction of fish species into European fresh waters: Economic successes or ecological disasters? *Bulleting Francais de la Peche et de la Pisciculture* 344–45:57–77

De Juana E (2004) Changes in the conservation status of birds in Spain, years 1954 to 2004. *Ardeola* 51:19–50

Echeverria C, Coomes D, Salas J, Rey-Benayas JM, Lara A, Newton A (2006) Rapid deforestation and fragmentation of Chilean Temperate Forests. *Biological Conservation* 130(4):481–494

Ereaut G, Segnit N (2006) *Warm words: How are we telling the climate story and can we tell it better?* Institute for Public Policy Research, London

Fischer A, Young J (2007) Understanding mental constructs of biodiversity: implications for biodiversity management and conservation. *Biological Conservation* 136:271–282

FAO (1988) An interim report of the state of the forest resources in developing countries. Food and Agriculture Organization, Rome

FAO (2006) Global forest resources assessment 2005: Progress towards sustainable forest management. Food and Agriculture Organisation of the United Nations, Rome

Gibbons DW, Reid JB, Chapman RA (1993) The New Atlas of Breeding Birds in Britain and Ireland: 1988–1991. T & AD Poyser, London

Groombridge B (1992) *Global biodiversity: status of the earth's living resources*. Chapman & Hall, London

Gu WD, Heikkila R, Hanski I (2002) Estimating the consequences of habitat fragmentation on extinction risk in dynamic landscapes. *Landscape Ecology* 17(8):699–710

Hails RS (2005) Assessing the impact of genetically modified crops on agricultural biodiversity. *Minerva Biotecnologica* 17(1):13–20

Hanski I (1999) Habitat connectivity, habitat continuity, and metapopulations in dynamic landscapes. *Oikos* 87(2):209–219

Hill JK, Thomas CD, Fox R, Telfer MG, Willis SG, Asher J, Huntley B (2002) Responses of butterflies to twentieth century climate warming: implications for future ranges. *Proceedings of the Royal Society B-Biological Sciences* 269(1505):2163–2171

Holdich DM, Pockl M (2005) Roundtable session 2 – Does legislation work in protecting vulnerable species? *Bulletin Francais de la Peche et de la Pisciculture* 376–77:809–827

Hoekstra JM, Boucher TM, Ricketts TH, Roberts C (2005) Confronting a biome crisis: global disparities of habitat loss and protection. *Ecology Letters* 8:23–29

Hulme PE (2003) Biological invasions: winning the science battles but losing the conservation war? *Oryx* 37:178–193

IPCC (2001) *Climate change 2001: the Scientific Basis, contribution of the working group I to the third assessment report of the Intergovernmental Panel of Climate Change*. Cambridge University Press, Cambridge

IUCN (1994) *IUCN Red List Categories. Prepared by the IUCN Species Survival Commission*. IUCN, Gland, Switzerland and Cambridge, UK

IUCN (2001) *IUCN Red List Categories and Criteria: Version 3.1. IUCN Species Survival Commission*. IUCN, Gland, Switzerland and Cambridge, UK

IUCN (2006) *2006 IUCN Red List of Threatened Species*. http://www.iucnredlist.org. Accessed 4 May 2006

Jones C, Moss K, Sanders M (2005a) Diet of hedgehogs (*Erinaceus europaeus*) in the upper Waitaki Basin, New Zealand: Implications for conservation. *New Zealand Journal of Ecology* 29(1):29–35

Jones PS, Young J, Watt A (2005b) *Biodiversity Conflict Management: A report of the BIOFORUM project*. CEH Banchory, Scotland, UK

Karlsson B, van Dyck H (2005) Does habitat fragmentation affect temperature-related life-history traits? A laboratory test with a woodland butterfly. *Proceedings of the Royal Society B-Biological Sciences* 272(1569):1257–1263

Karlsson B, Wiklund C (2005) Butterfly life history and temperature adaptations; dry open habitats select for increased fecundity and longevity. *Journal of Animal Ecology* 74(1):99–104

King KA, Zaun BJ, Schotborgh HM, Hurt C (2003) DDE-induced eggshell thinning in white-faced ibis: A continuing problem in the western United States. *Southwestern Naturalist* 48(3):356–364

Kunisue T, Minh TB, Fukuda K, Watanabe M, Tanabe S, Titenko AM (2002) Seasonal variation of persistent organochlorine accumulation in birds from Lake Baikal, Russia, and the role of the South Asian region as a source of pollution for wintering migrants. *Environmental Science and Technology* 36(7):1396–1404

Koh LP, Dunn RR, Sodhi NS, Colwell RK, Proctor HC, Smith VS (2004) Species coextinctions and the biodiversity crisis. *Science* 305(5690):1632–1634

Lafferty K, Kuris A (1996) Biological control of marine pests. *Ecology* 77:1989–2000

Lawton JH (1995) The response of insects to environmental change. In: Harrington R, Stork NE (eds) *Insects in a Changing Environment*. Academic Press, London, pp 3–26

Leakey R, Lewin R (1995) The sixth extinction: Biodiversity and its survival. Doubleday, New York

Lei FM, Qu YH, Lu JL, Yin ZH (2003) Conservation of diversity and distribution patterns of endemic birds in China. *Biodiversity and Conservation* 12:239–54

Lomolino MV (2000) Ecology's most general, yet protean pattern: the species-area relationship. *Journal of Biogeography* 27:17–26

MacArthur RH, Wilson EO (1967) The theory of island biogeography. Princeton University Press, Princeton

Macdonald IAW, Loope LL, Usher MB, Hamann O (1989) Wildlife conservation and the invasion of nature reserved by introduced species: a global perspective. In: Drake et al (eds) Biological invasions: a global perspective. John Wiley & sons, Chichester, pp 215–255

Mace GM, Stuart SN (1994) Draft IUCN Red List Categories, Version 2.2. *Species* 21–22:13–24

Mawdsley NA, Stork NE (1995) Species extinctions in insects: ecological and biogeographical considerations. In: Harrington R, Stork NE (eds) *Insects in a Changing Environment*. Academic Press, London, pp 321–369

Millennium Ecosystem Assessment (2005) *Ecosystems and Human Well-being. General Synthesis*. Island Press, Washington, DC

Miller K, Allegretti MH, Johnson N, Jonsson B (1995) Measures for conservation and sustainable use of its components. In: Heywood VH (ed) *Global Biodiversity Assessment*. Cambridge University Press, Cambridge, pp 915–1061

Newton I, Wyllie I (1992) Recovery of a sparrowhawk population in relation to declining pesticide contamination. *Journal of Applied Ecology* 29(2): 476–484

Nowicki P, Young J, Watt ADW (2005) *The Ecosystem Approach applied to Spatial Planning*. A report of the BIOFORUM project: CEH Banchory, Scotland, UK

Parmesan C, Yohe G (2003) A globally coherent fingerprint of climate change impacts across natural systems. *Nature* 421:37–42

Pimentel D, Wilson C, McCullum C, Huang R, Dwen P, Flack J, Tran Q, Saltman T, Cliff B (1997) Economics and Environmental Benefits of Biodiversity. *BioScience* 47(11):747–757

Pimm SL (2002) The dodo went extinct (and other ecological myths). *Annals of the Missouri Botanical Garden* 89(2):190–198

Redpath SM, Thirgood SJ, Leckie FM (2001) Does supplementary feeding reduce predation of red grouse by hen harriers? *Journal of Applied Ecology* 38:1157–1168

Redpath SM, Arroyo BE, Leckie FM, Bacon P, Bayfield N, Gutierrez RJ, Thirgood SJ (2004) Using decision modeling with stakeholders to reduce human-wildlife conflict: a raptor-grouse case study. *Conservation Biology* 18(2):350–359

Reid WV (1992) How many species will there be? In: Whitmore TC, Sayer JA (eds) *Tropical Deforestation and Species Extinction*. Chapman and Hall, London, pp 55–73

Reid WV, Miller KR (1989) *Keeping Options Alive: the Scientific Basis for Conserving Biodiversity*. World Resources Institute, Washington, DC

Ricketts TH (2001) The matrix matters: Effective isolation in fragmented landscapes. *American Naturalist* 158(1):87–99

Rodriguez C, Torres R, Drummond H (2006) Eradicating introduced mammals from a forested tropical island. *Biological Conservation* 130(1):98–105

Root TL, Price JT, Hall KR, Schneider SH, Rosenzweig C, Pounds JA (2003) Fingerprints of global warming on wild animals and plants. *Nature* 421:57–60

Roubik DW (1995) Pollination of cultivated plants in the tropic. Food and Agriculture Organization, Rome. *FAO Agricultural Services Bulletin* 118:1–196

Sala OE, Chapin FS, Armesto JJ, Berlow E, Bloomfield J, Dirzo R, Huber-Sanwald E, Huenneke LF, Jackson RB, Kinzig A, Leemans R, Lodge DM, Mooney HA, Oesterheld M, Poff NL,

Sykes MT, Walker BH, Walker M, Wall DH (2000) Biodiversity - Global biodiversity scenarios for the year 2100. *Science* 287:1770–1774

Sairinen R, Viinikainen T, Kanninen V, Lindholm A (1999) *Future visions of Finnish environmental policy.* Gaudeamus, Helsinki

Schtickzelle N, Choutt J, Goffart P, Fichefet V, Baguette M (2005) Metapopulation dynamics and conservation of the marsh fritillary butterfly: Population viability analysis and management options for a critically endangered species in Western Europe. *Biological Conservation* 126(4):569–581

Singh LS (2002) The biodiversity crisis: A multifaceted review. *Current Science* 82: 638–647

Stanners D, Bourdeau P (1995) Europe's Environment: The Dobris Assessment. European Environment Agency, Copenhagen

Stohlgren TJ, Schnase JL (2006) Risk analysis for biological hazards: What we need to know about invasive species. *Risk Analysis* 26:163–173

Stoll-Kleemann S, O'Riordan T (2002) From participation to partnership in biodiversity protection: Experience from Germany and South Africa. *Society & Natural Resources* 15:161–177

Tchamba MN (1996) History and present status of the human/elephant conflict in the Waza-Logone region, Cameroon, West Africa. *Biological Conservation* 75:35–41

Thomas CD, Cameron A, Green RE, Bakkenes M, Beaumont LJ, Collingham YC, Erasmus BFN, de Siqueira MF, Grainger A, Hannah L, Hughes L, Huntley B, van Jaarsveld AS, Midgley GF, Miles L, Ortega-Huerta MA, Peterson AT, Phillips OL, Williams SE (2004) Extinction risk from climate change. *Nature* 427(6970):145–148

Thomas CD, Lennon JL (1999) Birds extend their ranges northwards. *Nature* 399(6733):213

Towns DR, Atkinson IAE, Daugherty CH (2006) Have the harmful effects of introduced rats on islands been exaggerated? *Biological Invasions* 8(4):863–891

Travis JMJ (2003) Climate change and habitat destruction: a deadly anthropogenic cocktail. *Proceedings of the Royal Society of London Series B- Biological Sciences* 270(1514):467–473

United Nations Environment Programme - World Conservation Monitoring Centre (UNEP-WCMC) (2003) World Database on Protected Areas (WDPA). Version 6. WCMC, Cambridge, U.K

Valkama J, Korpimaki E, Arroyo B, Beja P, Bretagnolle V, Bro E, Kenward R, Manosa S, Redpath SM, Thirgood S, Vinuela J (2005) Birds of prey as limiting factors of gamebird populations in Europe: a review. *Biological Reviews* 80(2):171–203

Van den Hove S (2000) Participatory approaches to environmental policy-making: the European Commission Climate Policy Process as a case study. *Ecological Economics* 33:457–472

Veeramani A, Jayson EA, Easa PS (1996) Man-wildlife conflict: cattle lifting and human casualties in Kerala. *Indian Forum* 122:897–902

Vickery JA, Watkinson AR, Sutherland WJ (1994) The Solution To The Brent Goose Problem – An Economic-Analysis. *Journal of Applied Ecology* 31:371–382

Vitousek PM, D'Antonio CM, Loope LL, Rejmanek M, Westbrooks R (1997) Introduced species: A significant component of human-caused global change. *New Zealand Journal of Ecology* 21(1):1–16

Vitterso J, Kaltenborn BP, Bjerke T (1998) Attachment to livestock and attitudes toward large carnivores among sheep farmers in Norway. *Anthrozoos* 11:210–217

Vos J (2000) Food habits and livestock depredation of two Iberian wolf packs in the north of Portugal. *Journal of Zoology* 251:457–462

Wadsworth RA, Collingham YC, Willis SG, Huntley B, Hulme PE (2000) Simulating the spread and management of alien riparian weeds: are they out of control? *Journal of Applied Ecology* Suppl 1 37:28–38

Wagner S (2000) *Privatwaldbewirtschaftung in Natura 2000 Gebieten. Beschränkungen und ihr finanzieller Ausgleich. AllgemeineForstZeitschrift/Der Wald* 20:1069–1070

Walther G-R, Post E, Convey P, Menzel A, Parmesan C, Beebee TJC, Fromentin J-M, Hoegh-Guldberg O, Bairlein F (2002) Ecological responses to recent climate change. *Nature* 416:389–395

Walther G-R, Berger S, Sykes MT (2005) An ecological "footprint" of climate change. *Proceedings of the Royal Society of London Series B- Biological Sciences* 272:1427–1432

Warren MS, Hill JK, Thomas JA, Asher J, Fox R, Huntley B, Roy DB, Telfer MG, Jeffcoate S, Harding P, Jeffcoate G, Willis SG, Greatorex-Davies JN, Moss D, Thomas CD (2001) Rapid response of British butterflies to opposing forces of climate and habitat change. *Nature* 414:65–69

Wenzel M, Schmitt T, Weitzel M, Seitz A (2006) The severe decline of butterflies on western German calcareous grasslands during the last 30 years: A conservation problem. *Biological Conservation* 128(4):542–552

Whitfield DP, McLeod DRA, Watson J, Fielding AH, Haworth PF (2003) The association of grouse moor in Scotland with the illegal use of poisons to control predators. *Biological Conservation* 114(2):157–163

Wienburg CL, Shore RF (2004) Factors influencing liver PCB concentrations in sparrowhawks (*Accipiter nisus*), kestrels (*Falco tinnunculus*) and herons (*Ardea cinerea*) in Britain. *Environmental Pollution* 132(1):41–50

Wiles GJ, Bart J, Beck RE, Aguon CF (2003) Impacts of the brown tree snake: Patterns of decline and species persistence in Guam's avifauna. *Conservation Biology* 17(5):1350–1360

Wilson EO (1984) Biophilia: The Human Bond with Other Species. Harvard University Press, Cambridge, Massachusetts

Young J, Watt A, Nowicki P, Alard D, Clitherow J, Henle K, Johnson R, Laczko E, McCracken D, Matouch S, Niemela J, Richards C (2005) Towards sustainable land use: identifying and managing the conflicts between human activities and biodiversity conservation in Europe. *Biodiversity and Conservation* 14:1641–1661

Zimmermann HG, Moran VC, Hoffmann JH (2001) The renowned cactus moth, *Cactoblastis cactorum* (Lepidoptera: Pyralidae): Its natural history and threat to native Opuntia floras in Mexico and the United States of America. *Florida Entomologist* 84(4):543–551

Chapter 3
Chemical Contaminants in Food

**Ingemar Pongratz, Katarina Pettersson, and
Malin Hedengran Faulds**

1 Why Are We Writing About This Book

In Western societies today, the issue of food consumption has evolved from a relatively short chain of trading between producer and consumer to a complex chain of different parties. Today, food consumption includes large-scale production, time-efficient handling, transport, and packaging of food. Before the different food items reach consumers, many different steps, persons, and industrial processes have occurred that in many respects may alter the composition of food. However, food is not only an industrial sector but a source of well-being, joy, and depending on the consumption pattern, a source of health or disease.

Food is a truly global product, everybody consumes food. While certain cultural sectors might display a preference for particular items, the bulk consumption of food is general and unites all different levels of society.

This generality is, thus, a major issue when dealing with food. As food consumption is global, the adverse health effects coupled to food consumption are also global. This fact suggests that the effects on consumers' health are global and, depending on the quality of dietary products, can be greatly enhanced or alternatively be a major source for diseases.

Currently, there is a great deal of interest in developing novel foodstuff with specific characteristics. These food items, commonly denominated functional foods, have been developed to fulfill certain beneficial criteria, for example, certain dairy products with bacterial supplements. The idea to support the health of consumers is therefore feasible and already in progress. However, what about the effects of other, unwanted chemicals present in food? What risk do they pose to consumers? The manner in which food is marketed today inevitably leads to the presence of non-food-related compounds in the diet. A clear example is packaging residues like

I. Pongratz (✉)
Karolinska Institutet, Stockholm, Sweden
e-mail: ingemar.pongratz@ki.se

P. Pechan et al., *Safe or Not Safe: Deciding What Risks to Accept
in Our Environment and Food*, DOI 10.1007/978-1-4419-7868-4_3,
© Springer Science+Business Media, LLC 2011

Bisphenol A (BPA) and other plasticizers. These compounds are present in plastic products used for wrapping food prior to consumption and leak in various degrees into the foodstuff for later ingestion by the consumers. Currently, it is not known which health risks, if any, this involuntary intake of packaging material components constitute. The levels of exposure are generally low, but it is important to remember that exposure of consumers to these compounds persists for a long time, during decades, and therefore any adverse health effects may appear late in life.

2 Chemical Contaminants and Expected Impact

In addition to its nutritional and energy value, food also contains unwanted products. Chemical residues from different stages in the production chain are present in the ultimate product and can potentially affect the health of consumers. For the general consumer, food represents a major route of exposure to a broad variety of chemical residues and environmental pollutants.

Recent surveys performed by the National Food Safety Authority in Sweden have estimated that over 90% intake of chemical contaminants and environmental pollutants occurs through food consumption. In addition, recent estimates suggest that over 80% of total body burden of contaminants will occur during the first 5 years of life, clearly demonstrating that exposure to chemical contaminants through the diet affects consumers from the earliest stages of life and onwards. This is a distressing scenario as exposure to chemical contaminants may affect humans at very sensitive stages during development. Furthermore, the effects may display an extended lag-period and not become apparent until later stages in life. In fact, this could potentially mean that exposure of small children to different chemical residues in food may predispose them to disease in adulthood. Furthermore, exposure to contaminants occurs already at fetal stages. For example, several studies have demonstrated that common bulk chemicals, such as Bisphenol A, commonly found in plastic products, readily crosses the blood–placenta barrier and reaches the embryo.

3 Why Is the Study of Health Effects Caused by Contaminants Important?

Many studies have coupled food consumption to different types of diseases or in, some cases, to protection from different diseases. For example, Asian populations seem to be protected from certain types of tumor diseases due to a high intake of soy products. The protective effect against tumors has been attributed to the high content of hormonally active phytoestrogens present in soy. Thus, it seems likely that the general health status of a population can be influenced through diet. There are also cases of the opposite, where dietary intake of certain compounds results in health hazards. An illuminating example is aflatoxin, a fungal toxin that forms in mold growing on certain foodstuffs, such as nuts. Intake of aflatoxin-contaminated food has been associated with a higher prevalence of liver cancer in certain regions. Currently, few scientific studies have attempted to link chemical contaminants in food to human diseases. Therefore, much of the experimental evidence linking chemical contaminants and endocrine-disrupting chemicals to disease conditions is based on studies performed with pure compounds and have not taken into account the potential impact of the food matrix itself. Any food item represents a complex mixture of numerous compounds. This complexity makes the scientific interpretation of exposure studies performed with food commodities very complex and difficult. Taken together, in most cases the potential harmful effects following exposure to chemical contaminants through food items are currently not well understood. The scientific challenge to identify these effects is, however, extreme. In essence, food is a highly complex mixture of a large number of compounds and the interplay between these compounds is difficult to characterize. Another difficulty lies in discriminating the effects stemming from natural food components from those of man-made compounds and to understand how this complex interplay will affect human health. A concrete example is wine, which is naturally rich in estrogenic compounds, such as lignans and other polyphenolic compounds. These compounds have been shown to act through receptor proteins that heavily influence human health and disease. Wine can, however, also be contaminated with man-made compounds, such as bisphenol A from plastic containers used to store wine but also with pesticide residues used to protect the grapes. These trace amounts of man-made chemical compounds also possess the ability to activate cellular pathways that potentially affect human health. Thus, it is extremely difficult to separate the effects from natural wine components and man-made components and the net effects on human health are hard to assess. Importantly, it has been suggested in epidemiological studies that different disease conditions, such as several types of cancer, can be coupled to food consumption patterns and therefore possibly to the presence of chemical contaminants in food.

4 Dioxin: Description and Origin

Dioxins are a group of polychlorinated aromatic chemicals with similar structural characteristics. There are 210 known different dioxin-like compounds, polychlorinated dibenzo-p-dioxins and dibenzofurans (PCDD/F), out of which 17 are considered toxic and harmful to human health. The most toxic is the 2,3,7,8-tetrachlorodibenzo-p-dioxin, TCDD, also known as Seveso dioxin (Fig. 3.1).

The commercial value of dioxins is minimal and thus, dioxins are not synthesized for commercial purposes, except in small quantities for research. Dioxins are formed as by-products in various industrial combustion processes and are therefore widely distributed in the environment. The major emission sources of dioxin are municipal and clinical waste incinerators operating at low temperatures, paper mills that utilize chlorine to bleach pulp, metal industries, domestic heating facilities and traffic. Further, naturally occurring phenomena, such as

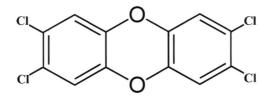

Fig. 3.1 Structure of TCDD

forest fires or volcano eruptions, contribute locally to increase the levels of dioxin contamination. A striking property of dioxins is an extreme persistence to biodegradation and also their fat-soluble ability, resulting in accumulation of dioxin and dioxin-like compounds in animal and human fat tissues for very long periods of time. For example, studies have suggested that the biological half-life of dioxin in humans is close to 20 years. Even though industrial release of dioxin and dioxin-related chemicals has dramatically decreased since the 1970s, concentrations in the environment are still high. Since dioxin is ubiquitous in the environment, humans are exposed through numerous ways with food being the major contributor for the general population.

4.1 Detecting and Assessing the Effects of Dioxins

The overall levels of dioxin in the environment or in most foodstuffs are relatively low, resulting in low-dose exposure for most populations. However, the presence of dioxin in the environment is always a cause of concern. Because of dioxins' high degree of persistence and ability to withstand biodegradation, accumulation in the food chain and low overall levels in the general environment suggest high levels at the top of the food chain.

Classically, detecting low levels of dioxin in food samples is a relatively straightforward but costly and work-intensive procedure. In particular, the sampling process is extensive and time consuming. Only recognized laboratories perform this arduous process, which includes a multitude of extraction and separation steps. An overview of the complex process is presented in Fig. 3.2.

To simplify the process, detection and quantification of dioxin and dioxin-like chemicals through other means have been attempted. In particular, large efforts have been made to develop biodetectors, such as cell-based systems utilizing readily detectable activity markers, so-called reporter cell lines, to assess the effects of dioxin in cells. The advantage of these systems is that they allow for detection of physiologically relevant responses to dioxin exposure. Disadvantages of the systems are that well-trained personnel and well-equipped laboratories are required and the technique cannot be used in field conditions (Fig. 3.3).

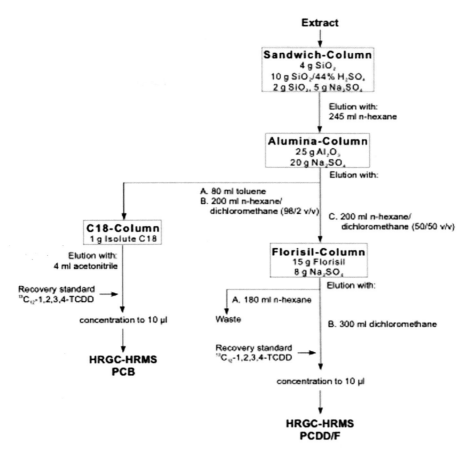

Fig. 3.2 Example of the lengthy process to detect dioxin in environmental samples

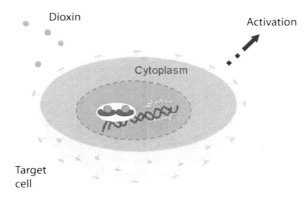

Fig. 3.3 A cellular system to detect dioxin exposure

4.2   Dioxin: Impact on Humans/Animals

TCDD gives rise to a wide range of toxic responses. These responses vary depending on the dose and, in many cases, on the type of animal species exposed. This observation suggests that genetic variations play a major role in dictating the harmful effects of dioxin. When exposed to high concentrations of dioxin, animals die from wasting syndrome within 2–6 weeks. Wasting syndrome is characterized by a dramatic loss of adipose tissue and a concomitant decrease in body weight. There are also reports stating that dioxin exposure causes damages to the thymus gland which, in turn, leads to alterations of the immune response, thereby interfering with the organism's ability to defend itself against infections. Dioxin is also known to induce damage to other vital organs, such as the liver, kidney, and the digestive tract.

Dioxin-exposed animals demonstrate severe reproductive disorders, including an elevated frequency of miscarriages and male sterility. Dioxin is also a potent teratogenic agent and TCDD exposure of fetuses has been shown to cause birth defects, including limb malformation and neurological effects. Dioxin can also act on a level where it affects other hormonal signaling pathways in the body, a phenomena known as endocrine disruption. For instance, the steroid hormone estrogen is the hormone responsible for proper development and function of the reproductive organs in both females and males. Dioxins have well-established antiestrogenic properties which impact the female reproductive tract, like inhibition of estradiol-induced increase in uterine wet weight, as well as decreased levels of estrogen receptors and progesterone receptors in rodent uterus (Safe and Krishnan 1995).

The variety of effects of dioxin exposure seen in most organs of the body may be attributed to interference with the regulation of cell proliferation. Growth is promoted by cell division and dysregulation of the timing of cell division induced by dioxin exposure can lead to severe effects. One example is the growing fetus where cell division disruption can result in birth defects, such as cleft palate. Another mechanism proposed for TCDD-action in the growing fetus is that exposure for a sustained period may produce signals at inappropriate times during organ formation. The result could be a delay in cell specialization, leading to various limb malformations. Dioxin causes uncontrolled growth of cells, suggesting a mechanism for cancer promotion. Other observed dioxin-related effects are on epithelial cells, which form the lining of many different organs in the body. The damage by TCDD on the thymus gland, urinary tract, liver, and bile ducts is caused by growth disruption of the epithelial cell population in these organs. For instance, chloracne is caused by hyper-proliferation and altered differentiation of epithelial skin cells.

The TCDD-dose that produces rapid death varies a great deal between species, with hamsters being least susceptible with a fatal dose of 5 mg/kg body weight while a dose 100 times smaller than this will kill a rat. Health-threatening effects, though, can be detected in far lower doses. As an example, mice given a dose of 4,000 times less than fatal dose will have significant less lymphocytes in the blood. These effects by TCDD on the immune system are similar for various animals. TCDD-induced changes in the thymus and liver occur at the same dose levels in hamster as in otherwise more sensitive species (Silbergeld and Gasiewicz 1989). Furthermore, animal data indicates the disruptive effects by TCDD on the fetus to be most critical for the developing immune system in proximity to birth (Rogan and Miller 1989).

4.3 Dioxin Effects in Humans

The major bulk of data available regarding the deleterious effects following dioxin exposure is derived from animal experiments. The effects of dioxin, however, vary considerable depending on the species exposed, where rodents appear to possess a substantially higher clearance ability of dioxin and dioxin-like compounds through biotransformation. This fact alone renders it difficult to extrapolate the results obtained from studies on laboratory rats or mice to human populations. It is well-established that humans are exposed to dioxin already from birth through breast feeding, and together with the extended half-life of dioxin in humans, one has to assume that health-hazardous effects of dioxin can result from a continuous accumulation, even when overall exposure levels are considered relatively low. Another obstacle in assessing or predicting health effects of dioxin and related compounds on human populations is that the physiological effects of dioxin are not well understood. Unfortunately, this gap of knowledge is not due to lack of examples of dioxin-exposed human subjects. Human exposure to dioxin has occurred following industrial accidents, like in Seveso, but also from Vietnam veterans involved in defoliation during the Vietnam war. Other exposure cases involve, for example, contaminated food like in Belgium and Spain. In Taiwan and Japan, children were subjected to dioxin through breast feeding by mothers who had consumed contaminated cooking oil. This resulted in impairment of the children's immune systems.

The dioxin family compounds also effects reproduction, both on the maternal and paternal sides. It has been suggested that reduced sperm count in industrialized countries can be connected with exposure to organochlorines (Stachel et al. 1989) and environmental chemicals may be blamed for half the cases of male infertility. Vietnam veterans, who were exposed to TCDD, show lower sperm count than the average population (Stachel et al. 1989). Research in Israel has suggested that infertile men have higher levels of PCBs and organochlorine pesticides in their blood than fertile controls (Pines et al. 1987). However, researchers found that organochlorine levels have steadily decreased over the years. Most of the men examined were born in the 1950s and 1960s, when pesticide and PCB use were at their height. They would have been subjected to high dioxin exposure before birth and as babies, a situation shown to cause sterility in animal experiments. Several childhood cancers have further been linked to mutations in sperm. Cells are highly prone to damage when dividing and new sperm are constantly produced by cell division. This is to be compared with egg cells that are formed in the womb before birth. Thus, DNA damage happens easier to sperm, although the effect may disappear when the toxic substance is removed.

Dioxin and dioxin-like chemicals have also been shown to be potent cancer promoters and to cause chloracne (Goldstein and Safe 1998; Silbergeld and Mattison 1987; Silbergeld and Gasiewicz 1989). Chloracne is a disease with persistent skin eruptions, often accompanied by severe disfiguration. Other symptoms of acute dioxin poisoning includes joint pain, headaches, fatigue, irritability, and chronic weakness. This syndrome can persist in the body for at least 30 years after initial exposure (Grandjean et al. 1991).

In general, humans appear to be less sensitive to the toxic effects of dioxin than many animals studied. The reason behind this observation may be due to differences between species. Alternatively, the reason could be due to the fact that humans consume less food per body mass. In addition, TCDD is stored in adipose tissue and therefore not being able to affect vulnerable organs. Further, the general weight increase among adult humans may protect them from the toxic effects of dioxin. Tissue sensitivity also seems to be lower than those seen in the most sensitive animals (Byard 1987).

Newborn children and human fetuses are, however, considerably more susceptive to the adverse effects of dioxin. Various different studies have confirmed that PCBs, dioxins, and furans are able to cross the placenta in humans (Koppe et al. 1992). The fetus susceptibility resembles that of more sensitive animals due to more rapid growth and cell division than at any time in later life. Furthermore, fetuses have not developed the important drug-metabolizing detoxifying system found after birth. Until near the time of birth, the fetus lacks fat deposits that might dilute the impact of toxic exposure and its small size suggests a large intake of contaminant per body weight. The blood–brain barrier is incompletely developed so vulnerability to central nervous system damage is increased (Jacobson et al. 1990). Pre-birth effects of the dioxin family include malformations, neurological effects, and changes to the immune system that might give rise to cancer or infections.

Also, a higher incident of miscarriage has been linked with higher levels of PCBs. A study made in Italy found that women, who had miscarried, had higher levels of PCBs in their bodies (Leoni et al. 1989). After the Seveso accident in 1976, exposed pregnant women had abortions because of fears of the effects and these fetuses had higher levels of chromosome aberrations than those from nonexposed women (Tenchini et al. 1983). In Vietnam, because the whole environment was contaminated with Agent Orange, both men and women were exposed to TCDD. Several effects on pregnancy are thought to be linked to dioxin exposure. One is hydatidiform mole, which is an abnormal growth of placental tissue in an uncontrolled way leaving the embryo undeveloped. In severe cases, there is an increased risk of cancer, which can be fatal without intensive chemotherapy.

A study of children, whose mothers had eaten PCB-contaminated fish from Lake Michigan in North America, was started in 1980 (Jacobson et al. 1990). An equivalent of two or three lake trout or salmon meals were ingested per month over a period of around 15 years before the children were born. The women participating in the Michigan study, suffered from direct PCB-related effects on their pregnancies (Swain 1991). Higher consumption of contaminated fish was associated with anemia before and during pregnancy, water retention and swelling and an increased rate of infections. Another serious condition, toxemia, has been associated with high blood serum levels of PCBs (Bercovici et al. 1980). This could be the effect of PCBs affecting the mother's immune system, resulting in a massive reaction toward the fetus characterized by high blood pressure and fits.

The children subjected to the highest exposure were found to have lower birth weight than the relatively nonexposed control infants. This was related to maternal fish consumption and to concentrations of PCBs present in umbilical cord serum. The highly exposed babies had smaller head circumference and were also born earlier (Swain 1991). Aside from the physical differences, the children were also examined for behavioral deficits. Almost half of the highly exposed children were classed as relatively unresponsive. Many of them also showed jerky uncoordinated and a greater number of abnormally weak reflexes. In a follow-up study at 7 months of age, infants with the highest exposure before birth showed impaired short-termed visual memory. The score on this test was shown to be dependent on PCB levels in umbilical cord serum. Interestingly, the researchers found that the test score was related only to pre-birth exposure to PCBs, the amount of PCBs received via nursing was indispensable. A second study was made on the children at 4 years of age. Children with higher intakes of PCBs before birth had lower scores on various types of short-term memory tests. There was no gross impairment, but definitively diminished potential.

The fact that damage of TCDD caused before birth was still present in children up to the age of four is alarming. It is possible that these effects could persist for life, and this theory is supported from a study made on rhesus monkey fetuses exposed to PCBs through their mothers (Schantz et al. 1989). The mothers were given a daily TCDD intake for 45–49 months and then a dioxin-free diet for 10–12 months before mating. TCDD was found to cause behavioral changes in the interaction between mother and child and possibly, decreased visual attention. The monkey babies born after TCDD exposure were subjected to a series of tests, similar to those carried out on the Michigan children. The most striking effect was the interaction between mother and baby, where TCDD-exposed infants spent more time close to their mothers and suckling. They ventured away from their mothers less, and when they did, their mothers fetched them back quicker than normal, a behavior often seen when babies are sick. These babies, however, showed no overt sign of illness. Other effects were also observed; the babies were more passive from birth and showed decreased visual attention at later stages. The results were rather similar to those found in the Michigan children. Taken together these studies indicate that effects on children are more likely to occur in populations living in areas with greater than average environmental dioxin burden and where fish consumption is also high or relatively high.

4.4 Sources of Dioxin Exposure

Everyone in industrialized countries is exposed to and has a mixture of dioxins, furans, co-planar PCBs, PCNs and other similar compounds stored and accumulated in their body fat. This chemical cocktail of compounds in our bodies is likely to add together, making up a total dioxin-like toxicity. The most toxic member of the dioxin family is TCDD with lethal effects at very low concentration. However, the reason for its potency is connected to the structural similarity to natural hormones. The power of hormones lies in the ability to act in minimal amounts as chemical messengers controlling vital processes in the body. Thus, an accidentally produced contaminant from the chemical industry, such as TCDD, can act as disruptors of the subtle messenger system in the body by disturbing or mimicking the action of hormones and thus disturbing the sensitive and fine-tuned endocrine system of higher organism. Dysregularities in the endocrine system has, in many cases, been shown to be closely coupled to impaired health and it is therefore not surprising that disruption may result in disease. A major scientific challenge, however, is to connect the effects of exposure to chemical contaminants, such as dioxin, to diseases that in many cases are manifested later in life, in many cases with a latency of decades.

4.5   Pollution/Accidents

As a result of a number of accidents and explosions in the 1940s, 1950s and 1960s, workers involved in the production of chemicals and herbicides were exposed to high levels of dioxins. The Seveso accident is a well-known example and in many cases served as a trigger for further dioxin research.

In July 1976, an explosion at a chemical plant in Seveso, Italy released a toxic cloud thought to contain around 30 kg of TCDD covering an area of about a square mile, affecting a population of more than 30,000 people. Reports of a higher rate of death from heart and chest diseases, a rise in cases of diabetes and an increased rate of certain uncommon cancers were noted. Furthermore, nearly 200 cases of severe chloracne were detected. Later studies also suggest that there is a change in the sex ratio of newborn children in the Seveso area with more male children being born.

Furthermore, during the Vietnam War in 1960s, the US army sprayed large quantities of herbicides over the Vietnamese countryside in an attempt to destroy crops and deprive the Vietnamese guerrilla of cover from tree foliage. They used a range of trichlorophenol herbicides, including the well-known Agent Orange, which was highly contaminated with dioxin. Later, during 1970s and 1980s, Vietnam veterans pursued legal claims against the US government claiming that following exposure to dioxins, the Vietnam veterans involved in spraying dioxin-contaminated Agent Orange displayed elevated frequencies in diverse illnesses, most seriously malignancies, sterility, and birth defects in children. From the Vietnamese side, there have been several reports on general health effects of the population due to the massive spraying.

4.6   General Pollution

Aside for accidental fires, uncontrolled backyard barrel burning and burning of animal carcasses are also considered to be a source of dioxin outlet (Lemieux et al. 2003). However, the UK government, who received harsh criticism regarding the handling of the dead farm animals in 2001, estimated that the burning of animals during the first 6 weeks of the foot-and-mouth crisis the same year have released 63 g of dioxins into the atmosphere while forest fires in Canada are estimated to release 60 kg of dioxins every year.

4.7   Food-Related Accidents

However, animals are not only exposed through grazing on contaminated pastures or through exposed hay/silage, some of the largest contamination incidents have involved contaminated animal feed.

A more recent case of dioxin contamination of food occurred in the southern part of the USA in 1997. Chickens, eggs, and fish were contaminated when a dioxin-containing ingredient was used in the production of animal feed. In Belgium 1999, poultry and egg products were reported to have exceptionally high PCDD/F values. This lead to the finding of severely PCB-contaminated chicken and pig feed stemming from waste vegetable oil. The same year, there was another report from Queensland, Australia, where they discovered that naturally occurring kaolin clay, used in the production of feed, caused severely increased dioxin levels in pigs. These incidents lead to restriction programs of PCDD/F contamination levels in raw materials, as well as in the finished animal feed product. As a result, dioxin contamination levels have been substantially reduced but only for animals under monitored feeding. High levels of dioxin have been shown in free-ranging poultry and egg, especially from smallholding farms, suggesting environment to be a substantial contributor of dioxin exposure (Table 3.1).

Food-related accidents involving dioxin have also occurred. For example, two separate accidents involving cooking oil contaminated with PCDD/F have occurred

Table 3.1 PCDD or PCDF and dioxin-like PCB in food products intended for consumer consumption. Adapted from WHO

Food item	Sum of dioxins and furans (WHO-PCDD/F-TEQ)	Sum of dioxins, furans, and dioxin-like PCBs (WHO-PCDD/F-PCB-TEQ)
Bovine animals, sheep	3.0 pg/g fat	4.5 pg/g fat
Poultry and farmed game	2.0 pg/g fat	4.0 pg/g fat
Pigs	1.0 pg/g fat	1.5 pg/g
Liver of terrestrial animals	6.0 pg/g fat	12.0 pg/g fat
Muscle meat of fish	4.0 pg/g fresh weight	8.0 pg/g fresh weight
Muscle meat of eel	4.0 pg/g fresh weight	12.0 pg/g fresh weight
Milk/milk products	3.0 pg/g fat	6.0 pg/g fat
Hen eggs/egg products	3.0 pg/g fat	6.0 pg/g fat
Animal fat	1.0–3.0 pg/g fat	1.5–4.5 pg/g fat
Vegetable oil and fat	0.75 pg/g fat	1.5 pg/g fat
Marine oil	2.0 pg/g fat	10.0 pg/g fat

in Japan, 1968 (known as Yusho incident) and in Taiwan 1979, termed Yuocheng. The affected population developed severe chloracne, hyperpigmentation and also showed signs of disturbed development (Schecter et al. 2006). Further, children were exposed to dioxin before birth through maternal diet (Tanabe 1988). These children showed a number of physical defects at birth, as well as, poorer perfor- mance on standardized intelligence tests performed in follow-up studies (Rogan and Miller 1989). Affected children were still being born 6 years after the incident. The children were of smaller size than normal with discolored skin and nails and abnormal teeth and gums. Many of the children also appeared apathetic and dull, and later, these children scored in the low range on intelligent tests. There were also high rates of infections, such as bronchitis (Rogan and Miller 1989).

4.8   The Baltic Sea

The Baltic Sea has long been an area of concern with high contamination levels of different PCBs. This affects all creatures living in or of the sea, like birds, seals, and fish (Bergman 1999). Although, contamination levels in Baltic wildlife have decreased steadily since the 1970s, fatty fish like salmon still contain significant levels. This will have serious impact on people with high consumption of local fish, for example, fishermen and their families. A recent study shows elevated PCB plasma levels in Swedish men, which was strongly correlated to their fish consump- tion (Sjödin et al. 2000). Other studies show a link between polluted fish from the Baltic and an increased risk of type 2 diabetes, increased risk of having children with lower birth weight, decreased semen function, increased risk of breast cancer, and osteoporotic fractures (Rignell-Hydbom et al. 2004; Axmon et al. 2004; Rylander et al. 1998; Wallin et al. 2004).

4.9   Routes of Dioxin Exposure

The primary source of exposure to TCDD and related congeners in human popula- tions is the diet. In fact, it is estimated that approximately 90% of the daily exposure to dioxin stems from food. Because of the fat-soluble properties of dioxins, fatty food of animal origin like dairy, meat, and fish products may contain a high TCDD content. There is a vast variation in daily dioxin intake due to different eating habits and food sources.

Although a diet rich in animal fat is considered the major dioxin source, veget- ables, and fruit grown in areas subjected to high dioxin outlets also contributes to total dioxin body burden. Agricultural conditions are of great impact of the dioxin content of the product. Contaminated soil is a big problem in areas close to big waste incinerators or industrial areas. Further, sewage sludge is used as fertilizer and shown to be rich in dioxin. There are several reports on smallholding farms close to chemical waste incinerators where free-ranging hens lay eggs with exceptionally high concentrations of dioxin. Generally, free-ranging animals, without controlled food intake, are reported to contribute to higher chemical waste exposure.

Other significant sources of TCDD contamination of food are due to the transfer or migration of dioxin from different packaging materials, especially cartons surrounding liquids. A classic example is the milk cartons of bleached paper but also other everyday utensils like coffee filters, cream and juice cartons, paper cups and plates, or microwave bags. In short, a large variety of regular household item have been shown to bleed dioxin into their content. However, due to increased knowledge and stricter restrictions, dioxin levels in packaging materials are decreasing steadily. Even though dioxin is a fat-soluble chemical, some pollution of drinking water occurs but this is not considered to be any major threat.

4.10   Food: Baby/Adult

Some experts claim, though, that the accumulated amount during lifetime occurs before birth. Dioxin readily passes the placenta and exposes the fetus to significantly higher levels than those of adults. Later on, the newborn child is exposed to high amounts of dioxin through nursing. Around 10% of lifetime exposure can occur via breast feeding, which translates to approximately 50 times the daily exposure of an adult.

PCDD/F contamination of human breast milk has been steadily declining since the 1980s. In the few countries that monitor breast milk on a regular basis, as much as a 50% reduction can be seen (LaKind et al. 2004). However, as the PCDD/F levels in serum lipids and the fat in milk are balanced, an approximate intake can be estimated. The levels of PCDD/F vary a great deal within and between nursing periods and decline with time postpartum. There are also other aspects of exposure through nursing, as the extent of breast feeding varies with socioeconomic status and ethnicity (Forste et al. 2001). The levels of contaminants between different areas are also varying, in both the industrial and developing part of the world. Some of the highest levels of dioxin in breast milk are found in agricultural areas of developing countries, known for extensive treatment of their crops with pesticides. There are reported that the Canadian Inuit women, whose main diet consist of marine food rich in fat, accumulate a heavy total body burden of persistent organic pollutants (Dewailly et al. 1994).

Even though infants have higher daily intake of persistent chemicals, the half-lives of these chemicals are considerably shorter in children compared to adults. The half-life of TCDD is around 6 months in an infant compared to 5 years half-life in the age of 40 (Michalek et al. 1996). This phenomenon is probably attributed to the rapid growth of adipose tissue volume in developing children and possibly also to increased fecal excretion observed in infants. Also noteworthy is that although TCDD body burdens increase during nursing, the levels do not remain elevated and do not exceed levels observed in adults. A comparison between breast-fed and bottle-fed infants shows that at 7 years of age, the total body burden of TCDD are similar in the two groups (Kreuzer et al. 1997). WHO recommends that breast milk should be the only source of food for the first 6 months of life and that the benefits from nursing outweigh the potential risks of chemical exposure.

Children's primary source of chemical exposure is food, just like in adults. On a body-weight basis, younger children, around 2 years of age have 30–40% higher intake than adults, older children in the age of 6 have about double chemical exposure risks. Teenagers have approximately the same chemical intake as an adult.

4.11 Mechanisms of Action of Dioxin and Dioxin-Like Chemicals

A changing environment is constantly surrounding all types of biological organisms from bacteria to eukaryotic cells. To survive, the cells need to be able to identify changes in their immediate surroundings and also to react and adapt their internal cellular functions to challenges imposed on them by the environment. These challenges include, for example, changes in nutrient composition and availability, exposure to harmful or even toxic compounds or changes in temperature to mention a few. To meet these challenges posed by a changing environment, cells specifically modify the activity of enzymes or proteins that perform specialized functions inside the cells. These functions include, for example, the ability to store nutrients when availability is high in the environment outside or alternatively to mobilize internal nutrient reserves if conditions are scarce.

A key factor in this adaptation process, however, is to correctly assess the conditions outside and following identification of the current environmental situation transmit this information to the inside of the cell. In addition, the cells need to adapt by changing activity of enzymes that are required to cope with current stress. Therefore, the cells need to recognize the challenge, transfer the information inside the cell and finally specifically adapt intracellular levels of selected protein to respond to the challenge.

Cells, adapt to a changing environment by altering the activity of enzymes. This alteration can be achieved either by changing the efficiency of the intracellular pool of proteins, so-called allosteric control of enzymatic activity or alternatively, by changing the total intracellular amounts of specific proteins. This latter level of adaptation is achieved mainly through regulation of gene expression, also known as transcriptional regulation.

Throughout evolution, biological organisms have developed numerous different systems to identify changing environment situations and subsequently to adapt to changes. One such system of recognition is through intracellular receptors.

Receptors are small proteins present in all cells throughout the body. Their function is to bind signaling molecules that cross the outer cellular membrane and then diffuse inside the cell. As these signal enter the cell they bind specifically to receptor proteins, forming a unit like a key in a lock. Examples of these signaling substances include steroid hormones, like estrogens and androgens and vitamins, like retinoids or vitamin D. Environmental pollutants, like TCDD, mimic the function of these signaling substances as they possess the ability to enter the cells and bind to a specific receptor. These compounds form a signaling unit through binding to a specific receptor protein. This unit can then influence the activity in the cell nucleus through binding to specific DNA sequences located in regulatory regions of inducible genes. By mimicking the action of endogenous signaling compound, endocrine disruptors trigger a cellular pathway at a time when this pathway should not be active and hereby interferes with the normal function of the cell. This outcome is commonly referred to as endocrine disruption.

The receptor that binds dioxin is called the aryl hydrocarbon receptor (AhR). The AhR is a member of a larger family of structurally related factors namely the bHLH–PAS family of transcription factors. The bHLH/PAS family comprises a prominent class of structurally related transcription regulators that control a variety of developmental and physiological events (for reviews, see, for example, Kleman et al. 1995; Dunlap 1999; Gu et al. 2000). Different proteins which belong to this family include the Sim transcription factor which regulates differentiation in the central nervous system protein (Crews et al. 1988; Nambu et al. 1991), tracheal development regulator Thr (Isaac and Andrew 1996; Wilk et al. 1996), circadian rhythmicity regulatory protein Clock (King et al. 1997), hypoxia-inducible transcription factor HIF-1α (Wang and Semenza 1995), AhR (Burbach et al. 1992; Dolwick et al. 1993), and ARNT (Hoffman et al. 1991).

4.12 Mechanism of Action of AhR

As stated before, dioxin is a very fat-soluble compound. When an organism is exposed to dioxin this compound easily enters into the systems through the skin as is delivered into the bloodstream.

In blood, dioxin is bound to fatty carrier proteins, like albumin, and is carried across the organism. However, dioxin can at any time be released and may freely cross the cell plasma membrane and enter the cell.

Inside the cell, dioxin will bind to the intracellular Ah receptor. This binding triggers a cascade of events, commonly referred to as the AhR-activation process. When dioxin is bound to the AhR and the receptor–dioxin complex migrates to the nucleus, p450 enzyme production is stimulated. This process is known as enzyme induction. The resulting enzyme harbors an activity specifically designed to break down xenobiotic compounds with a planar structure, like dioxin. However, since dioxin is highly persistent to biodegradation, AhR remains bound to DNA and p450 enzymes are continuously produced. This persistent activity has been reported in workers exposed to TCDD over 20 years previously, whose levels of liver enzymes are still abnormal. Further, elevated AHH activity has been detected in the placentas of pregnant women poisoned by PCB-contaminated rice oil 4–5 years earlier (Lucier et al. 1987). The biological significance of this persistent activation is currently not well characterized. The main function of p450 enzymes is to induce biodegradation and subsequent clearance of foreign compounds from the cell. During this clearing process, p450 enzymes produce chemically unstable intermediate metabolites that can lead to cell damage. A persistent activation of p450 enzymes by dioxin could therefore result in elevated concentration of damaging intermediate compounds inside the cells.

Although dioxin is not degraded by p450 enzymes, other substances are, including, for example, certain foreign chemicals and endogenous hormones. In many cases, these intermediate breakdown products are mutagens. The Ah receptor has functions other than enzyme induction, such as controlling certain cell functions, especially those regulating growth and differentiation of cells.

5 Managing Risk

5.1 The Toxic Equivalent Factors

In order to increase food safety, monitoring plans and risk assessment programs have been organized world-wide to evaluate the presence of contaminants in foodstuff and their potential impact on consumers' health. The "toxic equivalent factors" (TEF) concept is an attempt to estimate the potential health risks from total exposure to dioxin and dioxin-like chemicals (van den Berg et al. 1998). With this concept 2,3,7,8-TCDD is assigned a TEF of 1 and all other dioxin-like substances are evaluated relative to their toxic potency in comparison with TCDD. The TEF values are then used to calculate the toxic equivalent, a value that defines the total dioxin content of a sample. This measure of toxicity (TEF), however, cannot predict the exact effects of each dioxin-like compound on cell division or birth defects because these effects depend on the behavior of effected genes and this varies between different cells and species. It does not account for variations caused by the presence or absence of Ah receptors or for individual differences of ability to store pollutants in fat or to excrete them. Since humans are especially genetically diverse this is, in fact, a severe shortcoming. In addition, the effects of early exposure to dioxin (fetal or newborn stages) cannot be quantified using the TEF approach as the effects may only be visible later in life.

Dioxins and dioxin-like chemicals will add together to give a total toxicity. This has been shown experimentally for dioxins and furans (Eadon et al. 1986). However, a moderately toxic PCB which produced only mild pre-birth effects in mice, on its own was found to increase the TCDD effects in those mice tenfold (Birnbaum et al. 1985). However, new statistical methods of evaluating the precise contamination profiles from specific sources or products are being developed (Antignac et al. 2006) (Table 3.2).

There is extensive scientific evidence that exposure to dioxin and dioxin-like compounds are attached to considerable risks to human health. It is also important to state that the levels of dioxin in the environment today, with very specific exceptions, do not support the notion that dioxin exposure is accompanied with immediate, short-term threats to human health. In fact, besides from a few spectacular cases, like the poisoning of the Ukrainian president and suicide attempts, dioxin is not suspected to have caused immediate deaths. Looking at the levels of dioxin in the environment today and comparing these levels to the immediate toxicology of dioxin, the levels of dioxin are simply not high enough. However, the chemical properties of dioxin are very worrisome. Given the extreme stability displayed by dioxin, exposure to dioxin is highly cumulative. This means that low, daily levels of dioxin exposure will inevitable lead to a build-up of concentrations of dioxin to potential harmful levels.

However, today, there is a clear lack of scientific data regarding the risks that this fact may pose for consumers. In addition, it is very important to note that the bulk of experimental information is based on toxicological approaches that measure the effects of dioxin exposure to relatively high doses and short times of exposure. Given the fact that dioxin interferes with the endocrine balance and hormonal signaling, harmful effects following prolonged exposure to dioxin may not manifest themselves until long time after, even decades (Table 3.3).

This prolonged scenario poses extraordinary regulatory problems. Obviously, this situation is not specific for dioxin but includes the vast majority of chemical contaminants in food. The levels of contaminants that are currently present in our

Table 3.2 Exposure levels to dioxin in selected human populations (adapted from Swedish National Food Administration: Risk assessment of nondevelopmental health effects of polychlorinated dibenzo-p-dioxins, polychlorinated dibenzofurans, and dioxin-like polychlorinated biphenyls in food)

Cohort	Nr individuals	Levels in fat (ng/kg mean levels)
German chemical workers	1,189	110
US chemical workers	3,444	1,600
Seveso	80	94
Seveso	48	55
Seveso	281	66

Table 3.3 Examples of major accidents involving exposure to dioxin

Location	Year	Substance	Relative exposure levels	Observed effects (examples)
Seveso	1976	TCDD	Soil levels	Chloracne
			Zone A: >50 $\mu g/m^2$	Diabetes
			Zone B: 5–50 $\mu g/m^2$	Modification of birth sex ratio (increase F)
Yusho	1968	TCDD in rice oil	2,000/3,000 mg/kg	Chloracne
				Tooth deformities
Times Beach	1971	Oil spray		Animal death
Belgium	1999	TCDD in animal feed		Emergency animal sacrifice to avoid human consumption

diet are low, so the threat is through interference with hormonal balance. Clearly, the fact that imbalance in an organisms hormonal systems may lead to long-time disease after exposure makes regulatory measures difficult.

However, the current situation is clearly not satisfactory. The detrimental effects of dioxin, in particular, and other chemical contaminants on the hormonal balance raises the highly discomforting possibility that dioxin and other chemical residues in food may play a role in a number of human disease conditions. Numerous experiments have demonstrated, for example, that estrogen hormonal signaling is involved in regulation of body metabolism and lack of certain aspects of E_2-signaling pathways are coupled to obesity.

Dioxin is a potent antiestrogenic compound. It is therefore possible that exposure to dioxin may lead to alterations in metabolism and in the long run to obesity. This is a very discomforting scenario, both from a human aspect and from a social aspect, as it includes both a great deal of human suffering and high economic costs. It is, however, clear that in order to protect the consumers, high costs will be involved. An example is the fishing situation in the Baltic Sea. Currently, maximum levels of dioxins allowed in fish are 4 pg/g of fresh weight. The fish from the northern parts of the Baltic Sea display considerable higher levels, so a permanent adoption of this regulation effectively will lead to disruption of the fishing industry around the Baltic Sea. On the other side of the scale are the risks to human health following exposure to dioxins as a result of consumption of Baltic sea fish. In addition, there is an ethical aspect of informing consumers of this risk. Clearly, this is a very difficult situation to address.

6   Future Perspectives

Many of the chemicals that eventually end up in our diet stems from contamination by industrial chemicals and industrial processes. Recent estimates suggest that European consumers are exposed to 70,000 chemicals every day. The figure regarding exposure to chemicals through diet is currently not known but certainly a major part of the everyday exposure occurs through food. It is also relevant to mention that most likely, the majority of chemicals that we are exposed to on a daily basis, do not pose any threat to consumers health. Nevertheless, a safe estimate is that among this large number of compounds some are bound to pose a risk to consumers. As chemicals are part of daily life it is very difficult to imagine life without them. An example is packing material. How would we be able to transport food if it is not contained in a transport container? Lack of scientific knowledge about potentially harmful chemicals is hampering the decision making process, providing risk assessor and decision makers with a major challenge of developing appropriate chemical legislations. In a short-term scenario, we believe that more scientific information is required to allow legislators to make relevant well-informed decisions. However, equally important is development of new methods to interact with the public and new means to spread scientific information. In modern Western societies, consumers are constantly surrounded by different, sometimes conflicting, messages. Under these conditions, it is important to develop new methods to reach out to consumers with information regarding potential risks but also benefits coupled with food consumption.

7 Chapter Summary

Food production is a complex process that requires many steps. In a globalized economy, food components may originate from different parts of the world. They, in turn, may be processed in one country but sold somewhere else. Contaminants may enter the food product long this whole food processing chain. The study of contaminants is crucial because of their impact on human health and on the environment. There are many known contaminants. Yet it is often difficult to ascertain their mode of action. As the example of dioxins illustrates, contaminants are also affected by other chemicals. The outcome of this complex interplay is often difficult to predict and carries with it various risks. It is the task of scientists and risk managers to understand and assess these risks so that adequate safety levels can be established.

Glossary

Bisphenol A: A chemical often used in different plastics

Chemical Contaminant: Often referred to as trace amounts of man-made industrial or residual chemicals present in food

Dioxin: Highly toxic Environmental pollutant

Furan: Closely related to TCDD with similar properties

Nuclear receptors: Proteins that mediate the biological effects of hormones

Plasticizer: A chemical used in plastic

Teratogenic: A chemical compound capable of inducing birth defects

References

Antignac JP, Marchand P, Gade C, Matayron G, Qannari el M, Le Bizec B, Andre F (2006) Studying variations in the PCDD/PCDF profile across various food products using multivariate statistical analysis. *Anal Bioanal Chem* 384:271–279

Axmon A, Rylander L, Stromberg U, Hagmar L (2004) Altered menstrual cycles in women with a high dietary intake of persistent organochlorine compounds. *Chemosphere* 56:813–819

Bercovici B, Wassermann M, Wassermann D, Cucos S, Ron M (1980) Missed abortions and some organochlorine compounds: organochlorine insecticides (OCI) and polychlorinated biphenyls (PCBs). *Acta Med Leg Soc (Liege)* 30:177–185

Bergman A (1999) Health condition of the Baltic grey seal (Halichoerus grypus) during two decades. Gynaecological health improvement but increased prevalence of colonic ulcers. *APMIS* 107:270–282

Birnbaum LS, Weber H, Harris MW, Lamb JC,, McKinney JD (1985) Toxic interaction of specific polychlorinated biphenyls and 2,3,7,8-tetrachlorodibenzo-p-dioxin: increased incidence of cleft palate in mice. *Toxicol Appl Pharmacol* 77:292–302

Burbach KM, Poland A, Bradfield CA (1992) Cloning of the Ah-receptor cDNA reveals a distinctive ligand-activated transcription factor. *Proc Natl Acad Sci USA* 89:8185–8189.

Byard JL (1987) The toxicological significance of 2,3,7,8-tetrachlorodibenzo-p-dioxin and related compounds in human adipose tissue. *J Toxicol Environ Health* 22:381–403

Crews ST, Thomas JB, & Googman CS (1988) The drosophila single-minded gene encodes a nuclear protein with sequence similarity to the per gene product. *Cell* 52(1):143–151.

Dewailly E, Ryan JJ, Laliberte C, Bruneau S, Weber JP, Gingras S, Carrier G (1994) Exposure of remote maritime populations to coplanar PCBs. *Environ Health Perspect* 102 Suppl 1:205–209

Dolwick KM, Swanson HI, & Bradfield CA (1993) In vitro analysis of Ah receptor domains involved in ligand-activated DNA recognition. *Proc Natl Acad Sci USA* 90:8566–8570

Dunlap JC (1999) Molecular basis for circadian clocks. *Cell* 96(2):271–290

Eadon G, Kaminsky L, Silkworth J, Aldous K, Hilker D, O'Keefe P, Smith R, Gierthy J, Hawley J, Kim N, DeCaprio A (1986) Calculation of 2,3,7,8-TCDD equivalent concentrations of complex environmental contaminant mixtures. *Environ Health Perspect* 70:221–227

Forste R, Weiss J, Lippincott E (2001) The decision to breastfeed in the United States: does race matter? *Pediatrics* 108:291–296

Goldstein LS and Safe (1998) Tumors and DNA adducts in mice exposed to benzo[a]pyrene and coal tars: implications for risk assessment. *Environ Health Perspect* 106 Suppl 6:1325–1330

Grandjean P, Sandoe SH, Kimbrough RD (1991) Non-specificity of clinical signs and symptoms caused by environmental chemicals. *Hum Exp Toxicol* 10:167–173

Gu Y-Z, Hogenesch JB and Bradfield CA (2000) The PAS superfamily: sensors of environmental and developmental signals. *Annu Rev Pharmacol Toxicol* 40:519–561

Hoffman EC, Reyes H, Chu FF, Sander F, Conley LH, Brooks BA, Hankinson O (1991) Cloning of a factor required for activity of the Ah (dioxin) receptor. *Science* 252:954–958

Isaac DD, Andrew DJ (1996) Tubulogenesis in Drosophila: a requirement for the trachealess gene product. *Genes Dev* 10:103–117

Jacobson JL, Jacobson SW, Humphrey HE (1990) Effects of in utero exposure to polychlorinated biphenyls and related contaminants on cognitive functioning in young children. *J Pediatr* 116:38–45

King DP, Zhao Y, Sangoram AM, Wilsbacher LD, Tanaka M, Antoch MP, Steeves TD, Vitaterna MH, Kornhauser JM, Lowrey PL, Turek FW, Takahashi JS (1997) Positional cloning of the mouse circadian clock gene. *Cell* 89:641–653

Kleman MI, Overvik E, Poellinger L and Gustafsson JA (1995) Induction of cytochrome P4501A isozymes by heterocyclic amines and other food-derived compounds. *Princess Takamatsu Symp* 23:163–171

Koppe JG, Olie K, van Wijnen J (1992) Placental transport of dioxins from mother to fetus. II. PCBs, dioxins and furans and vitamin K metabolism. *Dev Pharmacol Ther* 18:9–13

Kreuzer PE, Csanady GA, Baur C, Kessler W, Papke O, Greim H, Filser JG (1997) 2,3,7,8-Tetrachlorodibenzo-p-dioxin (TCDD) and congeners in infants. A toxicokinetic model of human lifetime body burden by TCDD with special emphasis on its uptake by nutrition. *Arch Toxicol* 71:383–400

LaKind JS, Amina Wilkins A, Berlin CM Jr (2004) Environmental chemicals in human milk: a review of levels, infant exposures and health, and guidance for future research. *Toxicol Appl Pharmacol* 198:184–208

Lemieux PM, Gullett BK, Lutes CC, Winterrowd CK, Winters DL (2003) Variables affecting emissions of PCDD/Fs from uncontrolled combustion of household waste in barrels. *J Air Waste Manag Assoc* 53:523–531

Leoni V, Fabiani L, Marinelli G, Puccetti G, Tarsitani GF, De Carolis A, Vescia N, Morini A, Aleandri V, Pozzi V and et al (1989) PCB and other organochlorine compounds in blood of women with or without miscarriage: a hypothesis of correlation. *Ecotoxicol Environ Saf* 17:1–11

Lucier GW, Nelson KG, Everson RB, Wong TK, Philpot RM, Tiernan T, Taylor M, Sunahara GI (1987) Placental markers of human exposure to polychlorinated biphenyls and polychlorinated dibenzofurans. *Environ Health Perspect* 76:79–87

Michalek JE, Pirkle JL, Caudill SP, Tripathi RC, Patterson DG Jr, Needham LL (1996) Pharmacokinetics of TCDD in veterans of Operation Ranch Hand: 10-year follow-up. *J Toxicol Environ Health* 47:209–220

Nambu JR, Lewis JO, Wharton KA, Jr. and Crews ST (1991) The Drosophila single-minded gene encodes a helix-loop-helix protein that acts as a master regulator of CNS midline development. *Cell* 67:1157–1167

Pines A, Cucos S, Ever-Handani P, Ron M (1987) Some organochlorine insecticide and polychlorinated biphenyl blood residues in infertile males in the general Israeli population of the middle 1980's. *Arch Environ Contam Toxicol* 16:587–597

Rignell-Hydbom A, Rylander L, Giwercman A, Jonsson BA, Nilsson-Ehle P, Hagmar L (2004) Exposure to CB-153 and p,p'-DDE and male reproductive function. *Hum Reprod* 19:2066–2075

Rogan WJ, Miller RW (1989) Prenatal exposure to polychlorinated biphenyls. *Lancet* 2(8673):1216

Rylander L, Stromberg U, Dyremark E, Ostman C, Nilsson-Ehle P, Hagmar L (1998) Polychlorinated biphenyls in blood plasma among Swedish female fish consumers in relation to low birth weight. *Am J Epidemiol* 147:493–502

Safe S and Krishnan V (1995) Chlorinated hydrocarbons: estrogens and antiestrogens. *Toxicol Lett* 82–83:731–736

Schantz SL, Levin ED, Bowman RE, Heironimus MP, Laughlin NK (1989) Effects of perinatal PCB exposure on discrimination-reversal learning in monkeys. *Neurotoxicol Teratol* 11:243–250

Schecter A, Quynh HT, Papke O, Tung KC, Constable JD (2006) Agent Orange, dioxins, and other chemicals of concern in Vietnam: update 2006. *J Occup Environ Med* 48:408–413

Silbergeld EK, Gasiewicz TA (1989) Dioxins and the Ah receptor. *Am J Ind Med* 16:455–474

Silbergeld EK, Mattison DR (1987) Experimental and clinical studies on the reproductive toxicology of 2,3,7,8-tetrachlorodibenzo-p-dioxin. *Am J Ind Med* 11:131–144

Sjödin A, Hagmar L, Klasson-Wehler E, Björk J, Bergman A (2000) Influence of the consumption of fatty Baltic Sea fish on plasma levels of halogenated environmental contaminants in Latvian and Swedish men. *Environ Health Perspect* 108:1035–1041

Stachel B, Dougherty RC, Lahl U, Schlosser M, Zeschmar B (1989) Toxic environmental chemicals in human semen: analytical method and case studies. *Andrologia* 21:282–291

Swain WR (1991) Effects of organochlorine chemicals on the reproductive outcome of humans who consumed contaminated Great Lakes fish: an epidemiologic consideration. *J Toxicol Environ Health* 33:587–639

Tanabe S (1988) PCB problems in the future: foresight from current knowledge. *Environ Pollut* 50:5–28

Tenchini ML, Crimaudo C, Pacchetti G, Mottura A, Agosti S, De Carli L (1983) A comparative cytogenetic study on cases of induced abortions in TCDD-exposed and nonexposed women. *Environ Mutagen* 5:73–85

Van den Berg M, Birnbaum L, Bosveld AT, Brunstrom B, Cook P, Feeley M, Giesy JP, Hanberg A, Hasegawa R, Kennedy SW, Kubiak T, Larsen JC, van Leeuwen FX, Liem AK, Nolt C, Peterson RE, Poellinger L, Safe S, Schrenk D, Tillitt D, Tysklind M, Younes M, Waern F, Zacharewski T (1998) Toxic equivalency factors (TEFs) for PCBs, PCDDs, PCDFs for humans and wildlife. *Environ Health Perspect* 106:775–792

Wallin E, Rylander L, Hagmar L (2004) Exposure to persistent organochlorine compounds through fish consumption and the incidence of osteoporotic fractures. *Scand J Work Environ Health* 30(1):30–35

Wang GL & Semenza GL (1995) Purification and characterization of hypoxia-inducible factor 1. *J Biol Chem* 270(3):1230-1237 *Environ Health* 30:30–35

Wilk R, Weizman I, & Shilo BZ (1996) Trachealess encodes a bHLH-PAS protein that is an inducer of tracheal cell fates in Drosophila. *Genes & Dev.* 10:93–102

Chapter 4
The Food Choices We Make

Paul Pechan

1 Why Am I Writing About This Topic?

I had spent over 20 years trying to understand why plant cells divide and how they do so. Some of this research involved isolating the genes from one plant and placing them into another. In the mid to late 1980s our experiments were conducted in laboratories, growth chambers, or greenhouses. Generally, what we did see of plants were single cells through a microscope. We were pleased to see that the cells divided and developed into plants in special growth chambers and, perhaps later be tested in the field. But the potential release of such plants into the environment raised concerns amongst the public. They worried that once released, such artificially engineered plants could negatively impact our ecosystem, plant diversity, and may even be dangerous if eaten. To protesters outside the institute we became monsters instead of scientists, the creators of Frankenstein food or "Franken food."

Like many of my colleagues did, I spoke with the protesters. We wanted to let them see that our work was based on curiosity and the wish to understand the way things work and to ultimately help our society. We wanted to explain that what we were doing was not dangerous and ultimately to the benefit of mankind. To a large degree it was a failure not only because we neglected to see that the discussion was not only about facts, but also about attitudes to life, mistrust and broader political and ideological issues such as the power of large companies, monopolies, and life style choices. When the first transgenic plants came onto the market, it became clear that the conflicts concerning genetically modified crops or GM foods, as they became known, remained. Now, more than 25 years after the first report of on a transgenic plant, it is fitting to look back and review GM crops and their products. With the benefit of the hindsight, what can we now say about the benefits and risks of GM crops and their products? Are they harmful to our food chain?

The chapter is divided into a number of sections, each of them dealing with different aspects of the GM food debate.

P. Pechan (✉)
Institute of Communication and Media Research,
Ludwig Maximilians University Munich, Oettingenstr 67, 80538 Munich, Germany
e-mail: pechan@ifkw.lmu.de

P. Pechan et al., *Safe or Not Safe: Deciding What Risks to Accept
in Our Environment and Food*, DOI 10.1007/978-1-4419-7868-4_4,
© Springer Science+Business Media, LLC 2011

2   Why Have Genetically Modified Food?
Why Is It Important for the World?

GM crops and food are a consequence of scientific and technical advances that took place especially in the 1960s and 1970s. Many are happy with these developments – others would prefer to go back to a time without GMOs. So how did we get to the point where we are today, and was it a good decision to create GM crops and products?

2.1   History: The Big Picture

When discussing GM crops and products, the general public usually thinks of genetic engineering, where scientists take a gene from an organism (animal, plant, or a microbe) and somehow introduce it into a plant. However, the creation of GM crops was a result of scientific advances in many disciplines over many decades. The general umbrella name for these disciplines is biotechnology. Biotechnology can be broadly defined as application of biology and its processes, including the field of genetic engineering, to our everyday lives. Biotechnology's main aim is to alter the living world to meet the needs of the human race.

Already thousands of years ago our ancestors had knowledge of how to use biological processes such as fermentation or salting of meat to prepare new products. Brewing alcohol, making cheese, baking bread or curing meat was made possible using these natural processes. They are all typical examples of how our ancestors manipulated and modified what nature had to offer in order to meet their specific needs. Domestication of plants and animals, through the use of various breeding methods, is another example of modifying nature. Indeed, traditional plant and animal breeding methods can be defined as a branch of biotechnology because plants and animals were and still are being modified through selection for improved characteristics. One just needs to look at the various types of dogs: large, small, bred to hunt, or protect – they are all a product of a deliberate human modification. The domestication of wheat, corn, or potatoes are additional excellent examples of breeding activities. Initially, the breeding activities were based on criteria that one could see (visible phenotypic characteristics) or that one could measure. Nevertheless, they also had the effect of modifying the genetic makeup (the genotype) of the plant. Indeed, all domesticated animals and plants have, to some degree been genetically modified through the repeated selection process to meet the needs of mankind. However, it is only over the last 100 years that traditional plant breeding has made rapid progress. This is mainly because of advances in genetics, the study of how genes are inherited and how they affect the characteristics of living organisms.

Initially, what these biotechnological activities had in common was a need to produce or protect our food. In addition, agriculture also increased the stability, wealth, and importance of the particular nation or region.

Access to sufficient and nutritious food is a basic right of every human being. It is not possible to talk of progress when people have insufficient food to eat. It is easy to forget that, until recently, famine was an ever present threat in many industrialized countries. And, it is still affecting millions of people in developing countries. We have no option when it comes to the modification of nature – we have to do it to survive. The so-called Green Revolution during the 1960s, based on new and improved crops that could be grown in combination with fertilizers and pesticides, greatly increased crop productivity. For the first time, many developing countries could claim to be self-sufficient in food production and provide adequate food for their populations.

Today, we are producing too much food in one part of the world, and not enough elsewhere. The question is whether the way we are continuing to technologically modify nature today to meet our nutritional needs may create new challenges and risks that we will need to deal with in the future. There may be other non-biotechnology options to feed the needy.

Indeed, with the new biotechnology techniques and tools at our disposal, production of more food and healthier food is no longer our only objective. We can now use plants to make specific products. For example, they can be used for specific purposes – such as enhancing vitamin contents or they can be made to be delivery systems for vaccines. These so-called functional foods are now entering our markets.

Today, biotechnology has many applications. It is used, for example, in medicine as well as in food production. Biotechnology allows for the development of products to treat human diseases. Its products can contribute to the treatment of cardiovascular diseases, cancer, stroke, etc. Bacterial cells can produce human proteins such as insulin or artificial transplant materials such as blood vessels or skin tissue. Biotechnology applied to agriculture had been given the name of plant genetic engineering. It allows the modification of life forms, their processes and products to an extent not possible before, enabling the targeted transfer of individual properties in the form of only one or a few desirable genes from one organism to another. In the public arena, genetically engineered plants have been called genetically modified, or GM plants. GM plants have a wide range of applications. They can be used, for example, to solve agriculture-related problems or to produce edible vaccines for infectious diseases such as cholera or hepatitis B. Such applications may be especially important in the developing world.

3   GM Crops and Food

3.1   *When First Grown?*

GM crops have been first planted in the open in the mid-1980s. These were experimental fields, whose purpose was to allow scientists to evaluate the possible impact of GM crops on the environment. The first commercially produced GM crop, a tomato containing a gene that slowed down its maturation process (FLAVR SAVR®), was approved for human consumption in the USA in 1994 and was introduced onto the market in 1996. Although not a great commercial success, it was nevertheless the first in a long list of GM crops now on the market.

3.2   How Much Is Being Grown and Where?

The land area under GM crop cultivation has been steadily expanding. The total area under GM crop cultivation has now passed 110 million hectares. This is nearly two times as much as it was beginning of the new century. Although it occupies a large area, it is still representative of less than 15% of all cultivated land worldwide, that being 700 million hectares (Table 4.1).

The main GM crops, in the descending order of priority, are as follows: soybean, maize, cotton, and canola (rapeseed). There are other crops that have been genetically engineered including, for example, tomatoes and potatoes. GM soybean dominates the market because of the crop importance in the USA and South America. As soybean is an important component of many processed foods, directly or indirectly GM soybean products find their way to consumers. GM crops have been developed mainly to resist insect attacks and to be tolerant of chemical sprays that are used to kill weeds.

Interestingly, the countries that led in the planting of GM crops 10 years ago are still the leaders in this field. USA, Argentina, Brazil, and Canada head the list. These countries grow more than 90% of the world's GM crops. However, it is likely that a number of Asian countries will become major GM crop growers, especially China and India. This may not be surprising as both countries have major problems with insect damage to crops. Both are expanding their GM cotton production. Rice may soon become a major GM crop. Europe is a minor grower of GM crops: it grows less than 0.3% of the world's GM crops. This is primarily because of the past de facto moratorium imposed on GM crops in Europe until the year 2003, current strict regulations and the refusal of the European consumers to buy GM products.

3.3   Why Grow Them?

GM crops were initially introduced onto the world markets to meet a specific need of farmers to allow better management of weeds and pests. This meant namely to reduce insect or viral damage to crops and to make spraying of weeds easier and more selective.

Table 4.1 GM crop production based on land area usage for the main 4 crops (ISAAA Briefs No 36-2007)

	Area	Area GM	Proportion GM (%)
Soybean	91	58.6	64
Maize	148	35.2	24
Cotton	35	15	43
Rapeseed	27	5.5	20

Thus these GM crops had distinct agronomical objectives. Although the first generation GM crops were initially marketed by the biotechnology companies as products that will significantly increase farmers income (by helping farmers to grow more and better quality crops), it became clear that such predictions were too optimistic. Although farmers may increase their crop yields (and thus increase their financial rewards), these returns are dependent on many variables, such as the timing of pest attacks, geographical location, time of the year, and weather conditions. What is clear is that at the very minimum, the first generation of GM crops reduces the exposure of farmers and farm animals to unnecessary chemicals and provide the farmer with the comfort of spraying weeds while not having to worry about damage to their crops.

The situation is entirely different with the second and third generation GM crops now under active development. Unlike the first generation of GM crops that benefited primarily the farmers and the industry, second and third generation GM crops provide concrete benefits, such as improved taste and quality, to the consumer. These crops will have enhanced nutritional, health, and improved functional value such as eliminating allergens from the food product. In addition, these new crops will also have the ability to adapt to extreme environmental conditions. GM crops will be used as bio-factories, producing, for example, edible vaccines.

However, as with all new technologies, there are doubts about the benefits of GM crops and indeed concerns about possible risks both to the human population and the environment. In order to start a more detailed discussion about the benefits and risks of GM crops, it is first useful to briefly summarize what it is about GM crops that makes them subject of such heated debate.

3.4   What Is Different About Creating GM Crops from New Conventional Crops?

In theory, scientists can potentially create almost an infinite number of genetically modified plants. The limiting factor is no longer the technology, but rather the ability to identify a suitable gene to be transferred into the recipient plant and to make sure that the gene becomes integrated and expressed in the new host.

Genetic modification of plants is what is termed a targeted approach to plant improvement. One takes one, two, or sometimes more genes of known function and transfers them into another plant where the benefits of these genes can be manifested. This is done asexually, by injecting or electroporating these new genes into recipient cells or shooting small pellets coated with the selected genes into the cells. This is a very targeted and controlled process. The use of these genes, or transgenes as they are sometimes called, should be a more predictable approach than traditional plant breeding.

This is in stark contrast to conventional plant breeding where thousands of unknown genes are exchanged between plants. Although conventional plant breeding also uses elements of genetic engineering, namely molecular markers to identify useful genes (traits), there is no isolation and transfer of individual genes between plants. Rather, what usually happens is that two plants that have interesting traits are sexually crossed (cross pollinated) so that the thousands of genes that make up the genome of the two plants are mixed. After subsequent screening, plants with the most desirable traits are selected. This is a rather uncontrolled process with many possible and unpredictable outcomes.

However, conventional plant breeding has been around much longer than the genetically modified approach. Moreover, conventional plant breeding relies on crossing and creating plants in a way that nature had done for millions of years. Unlike traditional plant breeding, genetic modification of plants is new and untested. It imposes on nature a new level of manipulation. We can now create new plant types much quicker than would happen in nature. And, with this new technology, we can do what nature can not: transfer genes between unrelated organisms.

3.5 How Safe?

Genetic modification of plants has made, perhaps understandably, many people uneasy. GM crops and products are new, man made and untested over time. Fear and mistrust color our attitude to GM crops: fear of unknown health and environmental risks and mistrust of our ability to properly identify and manage these risks.

There is a potential risk that GM crops and foods might contain allergic or toxic compounds. They should be therefore tested for possible allergic and toxic effects on living organisms. Some environmental groups are also concerned about the effect of GM crops on the environment. They are especially worried that, unlike the car, for example, it is not possible to recall them for repair if something goes wrong. Discussions about GM crops and food have taken on a much wider scope and now also include issues dealing with morality, choice, globalization and sustainability.

As GM crops need to be dealt with in an everyday environment, with its various interdependencies and uncertainties, it is often difficult for scientists to clearly elucidate many of the potential risks. This creates an interesting paradox. On the one hand, scientists acknowledge that there are limits to their knowledge and that there will likely remain some uncertainties about the impact of GM crops, especially on the environment. This uncertainty makes scientific arguments "soft." On the other hand, opinions about GM crops are varied and usually very passionate. These opinions become "hard." In many respects, in deciding about the usefulness and impact of GM crops, personal attitudes and perceptions dominate the "soft" scientific arguments. This further complicates the discussions about GM crops and their products.

However, before embarking on a more detailed risk analysis of GM crops on a scientific, social, political, and economic level, it is useful to remind ourselves what genetic engineering is really all about.

4 How Genetic Engineering Works

Genetic engineering of plants is very complex. It relies and takes advantage of many technical advances and scientific discoveries. Only when these were in place, was it possible to attempt to create genetically modified plants. Thus the technology on which genetic engineering of plants is based upon represents the accumulated knowledge from many scientific fields. Below is a summary of some of the key advances that made the creation of GM crops possible.

4.1 Understanding the Principles of Genetics

The ground rules for natural selection and genetics were established in the nineteenth century. Two men made major contributions to these advancements: Charles Darwin and Georg Mendel. Darwin with his publication of On the Origin of Species in 1859 and Mendel with his paper on basic mechanisms of heredity (inheritance of traits) in 1865. These advances helped scientists eventually to understand how plant characteristics are transferred and inherited from one plant to another – through genes.

4.2 The Role of DNA, RNA, and Proteins

Once the importance of genes was established, race was on to identify the gene structure. Thanks to the advancements in a number of scientific fields, such as chemistry and X-ray crystallography, scientists were able to identify DNA as the building blocks of genes. Finally, in 1953, Watson and Crick were able to describe the structure of the DNA – the famous double helix. A way was opened to start thinking about how the information encoded in the DNA was translated into specific cellular action. It was subsequently discovered that when DNA is activated, the message it contains is transcribed onto a messenger called RNA that transports the information from the cell nucleus to ribosomes in the cell cytoplasm. There, the information is translated into proteins. The role of proteins is to make things happen in the cell. A good example are enzymes. They are a type of a protein that mediates (catalyzes) reactions in the cell. Certain types of enzymes have turned out to be crucial for the manipulation of DNA – a key element of molecular biology and genetic engineering.

4.3 Enzymes and DNA Manipulation

One of the key discoveries that allowed the development of molecular biology was the observation that certain enzymes can cut both strands of DNA. These enzymes are called restriction enzymes. With the help of these enzymes it is possible to cut DNA at specific locations. The resulting DNA fragments could be then rejoined using another type of enzymes (called ligases). Indeed new pieces of DNA could be inserted between the DNA fragments and spliced again together. This type of DNA manipulation opened the way for inserting useful DNA into genomes of unrelated organisms (first success was in 1973). This new technology was called genetic engineering. Werner Arber, Daniel Nathans and Hamilton Smith received a Nobel Prize for their enzyme and DNA work in 1978. The ability to cut and re-join DNA in specific places opened up new possibilities to analyze DNA with methods such as Southern analysis or PCR (polymerase chain reaction), the latter being invented to multiply DNA sequences in vitro. These abilities, coupled with analysis of RNA and proteins, aptly named Northern and Western analysis, respectively, and DNA sequence analysis of animal and plant genomes made it possible to identify, evaluate, and later transfer useful genes from organism to organism. However, before this could be done, two more pieces of the puzzle needed to be in place. One was to be able to have suitable living cells where the genes of interest would be delivered. The second piece of the puzzle was to find out how to deliver DNA into living cells.

4.4 Living Cells: Plant Regeneration from Single Cells

Manipulation of plant cells is based on the knowledge how to culture plant tissue. This scientific discipline has a long history: already at the start of the twentieth century scientists were attempting to achieve regeneration of plants without the need of sexual crosses. This type of plant regeneration is called somatic (asexual) embryogenesis. It is based on the use of plant cells, chiefly clumps of undifferentiated cells (called callus) and protoplasts (plant cells without walls), that do not need gamete fusion. All somatic cells have a full complement of genes capable of forming a new plant from a single cell. This is achieved under highly defined in vitro culture conditions. The challenge was to define these growth conditions. As it turns out, many plant species have unique tissue culture requirements needed for regeneration of plants from single cells. Today, most of the economically important crops can be regenerated either from single cells or from callus.

4.5 DNA Delivery into Plant Cells

Once genes of interest have been isolated from the donor organism and the host plant cells cultured, what remained was to decide how to deliver the chosen gene into the plant cell. There are basically four methods to do this. The first is cocultivation of plant cells with Agrobacterium containing the genes of interest, the second is to deliver genes using a particle gun into plant tissue, the third method relies on electroporation of plant protoplasts and the fourth is delivery of DNA by microinjection. Most often used is the particle gun approach where small particles are coated with the desirable DNA and shot into the recipient cells. Agrobacterium cocultivation is also widely used. It relies on the fact that Agrobacterium (a type of bacteria) is able to penetrate plant cell walls and inject its DNA content into the cells. To efficiently deliver genes of interest, the DNA first needs to be inserted into a carrier, called plasmid. The plasmid is then introduced into the Agrobacterium which is then allowed to cocultivate with the host plant cells. The gene of interest is then integrated into the host genome. Once the cells have been transformed, they can be encouraged to grow into full-sized plants and the best selected for further propagation.

It is now a routine to isolate DNA from one organism, combine it with DNA from another (resulting in so-called recombinant DNA) and placed into cells of a third organism, giving it a new characteristic. The Bt potato, for example, contains a gene from a soil bacteria, *Bacillus thuringiensis,* that makes potatoes resistant to the Colorado potato beetle.

4.6   Development of Verification Methodologies

Genetically modified plants may create undesirable side effects. In order to reduce these risks, before the plants are released onto the market, a number of actions need to be taken. One of the main requirements is that GM crops can be monitored at any point from the farm to the market. A key element is the ability to detect, identify, and quantify DNA newly introduced into a plant or a food product. The aim is also to establish how much of the product is genetically modified in order to decide whether it complies with national GMO legislations.

4.7   Detection and Identification of GMO

Genetic modification usually involves insertion of a piece of DNA (the transgene) into the genome of the organism to be modified. Such newly inserted DNA sequences can be detected both at the DNA or protein level. DNA-based method is primarily based on multiplying a specific DNA with the polymerase chain reaction (PCR) technique. Two short pieces of synthetic DNA (called primers) are needed, each complementary to one end of the DNA to be multiplied. During the reaction, many copies of the target transgene are made and subsequently visualized using gel electrophoresis. When no copy is detected, transgenic DNA is not present. This method does not indicate whether the introduced DNA (gene) is active. This can, however, be established using the protein analysis method that measures the gene expression. It is based on detecting proteins with antibodies, usually against a specific protein, and with enzyme assays that also detect protein activity. This method cannot be used efficiently on processed food.

Detection and analysis of GM samples is useful only if both the positive and negative controls are available for comparison with the analyzed GM samples. This is the so-called certified reference material.

5 Risk Governance: What Do We Do to Make Sure GM Crops Are Safe?

As already explained in the introductory chapter on risk, risk governance is divided into three different phases: risk assessment, risk management, and risk communication. These phases overlap with each other, both in objectives and timing.

Risk analysis of GM crops and their products falls primarily into the category of uncertain and ambiguous risks. If there was, for example, a scientific clarity about the possible impact of GM crops on the environment, there would not be so much discussion about GM crops. It would have been a risk analysis (either falling within the category of simple or complex risks) where the cause and effect could be established with sufficient certainty. The problem with some GM crops is that there is a lot of uncertainty about their impact on our society and the environment. For that reason, many risk assessment studies need to be carried out on individual GM crops, before their general release, to reduce the uncertainty about the safety of the crop. Such studies allow decision makers to take management decisions about the conditions under with the GM crops can be marketed. Throughout this process, communication with the various stakeholders needs to take place to engage them in a dialogue about key safety issues.

The challenge faced by GM crops and their products is to adequately address the complexity of issues that they raise. GM crops are new, untested by time and fiercely contested by opponents. Risks associated with GM crops may be short or long term, immediate or delayed, and reversible or irreversible. There are also indirect environmental effects that occur, for example, through interaction with other living organisms. GM crops may have multiple impacts, not just at the level of health and environmental, but also at social, economic, legal, and political levels. These indirect risks are called systemic risks.

Application of genetic engineering to crops creates both a technology inherent risk as well as technology transcending risk systemic scenarios. The former is concerned with the direct impact of GM products on our health and the environment, the latter with the political and social context in which the technology is used and how these uses may benefit and/or harm the interests of different groups in society, taking into account other available technologies.

This complexity of risk analysis makes decisions about the safety of GM crops very difficult. The deliberations about GM crops are further complicated by the fact that GM crop safety issues are being considered within the broader frameworks of sustainability, globalization, and rights of individuals.

So how do decision makers deal with GM crops and their products? This, and the following two sections, explains the different phases of the GM crop risk governance process: risk assessment, management, and communication. Section eight then gives practical examples, in the form of a case study, how GM crops have been dealt with in Germany.

5.1 Risk Assessment

Because of the uncertainties about their direct health and environmental impacts, GM crops may contain potential risks. It is the objective of risk assessment to identify and evaluate these potential risks prior to releases into the environment and placing on the market. Risk assessment should also be carried out for all foods that have not been used for human consumption to a significant degree in the European Union before. It is important at this point to make a distinction between a hazard and risk.

Hazard can be defied as a potential source of danger or harm. Risk, as related to GM crops, can be in turn defined as unforeseen negative effects of a hazard. Risk can be expressed as the likelihood of an event occurring combined with the magnitude of the impact (how many people die from it or get ill?). These two elements are at the core of any risk assessment deliberation.

The first risk assessment step is establishment whether a potential hazard exists. This determines whether risk analysis needs to be carried out. For example, in the European Union, all novel foods, regardless whether they are GM or non-GM, are deemed to pose a potential hazard and are automatically subject to risk analysis. Every GM crop to be placed onto the market is assessed on a case-by-case basis and in a stepwise manner. For environmental release authorization this means that the specific GM crop is tested first in the laboratory then on a small scale in a field trial, followed by a large-scale field trial. At the same time, all GM crops need to be tested for potential risks to human health.

Once it is established that a risk analysis is required, a series of decisions need to be carried out about the criteria and methodology of the risk analysis. Criteria are the terms of reference used that frames the discussion determines the type of the obtained results. Criteria and methodology determine the quality and fairness of subsequent risk management decisions that precede introduction of a product onto the market.

The health and environmental criteria for GM crop risk analysis are summarized in Table 4.2. These serve to identify characteristics that may cause adverse effects. There are many possible methodologies to identify and analyze these characteristics. It is important that the methodologies used reflect the latest technological and scientific advances. As different analytical methodologies have their own advantages and weaknesses, there is a need to create common international standards. However, there are at present no agreed upon international criteria and methodology standards. Each country, or region, can impose its own set of criteria and methodology requirements before allowing planting or import of GM crops or their products. Although these requirements need to be fair and nondiscriminatory they may be of different scientific standards and may emphasize diverse criteria and methodologies (for more information, please see Sect. 4.2).

Table 4.2 Summary of general criteria for risk assessment of GM crops (from Pechan and de Vries (2005))

Expression of toxic or allergenic compounds
Potential for production of substances that are toxic or allergenic to human beings or other species
Other health risks
Antibiotic resistance, etc.
Effects on biogeochemistry
Potential to negatively influence decomposition processes in the soil and thus causing changes in nitrogen and carbon recycling
Increased persistence on the environment and invasiveness
Potential to confer an ecological fitness advantage to the GM crop causing persistence and invasiveness ("superweeds")
Transfer of genetic material
Potential to transfer the newly introduced genetic material to other crops or weeds via cross-pollination or to other organisms via horizontal gene transfer. Depending on the transferred trait such gene transfer might not present a hazard
Instability of genetic modification
Potential of reversing downregulation of a naturally occurring hazardous trait
Unintended effects
Potential that genetic modification leads to unintended effects, e.g., influencing other genes of the organisms, which might lead to unexpected hazards

What specific criteria or methodologies need to be looked at or used is a scientific and a political decision that reflects the influence of the various stakeholders. As already indicated, risk assessment of GM crops is done on step-by-step and case-by-case basis prior to the crop release onto the market. However, some countries or regions, as part of the risk assessment evaluation procedure, require additional long-term monitoring GM crops, even after they or their products are allowed onto the market. Long-term monitoring is part of the unintended effect analysis. Its aim is to identify unexpected side effects of GM crops so that, if shown to be of hazard to human health or the environment, they can be quickly identified and removed from the market. Thus monitoring of GM crops and their products concentrates on traceability efforts. As such the main requirement is that GM crops and their products can be detected at any point within the food chain, from the farm to the market. While detection and identification of GM raw material on the farm is relatively easy, in processed food the detection became more and more difficult indeed in some cases almost impossible (Table 4.3).

Detection and analysis of GM samples is useful only if certified positive and negative controls are available for comparison with the analyzed GM samples. This material is needed to allow laboratories to calibrate their equipment and procedures. These comparisons will greatly increase the confidence and ability to make meaningful conclusions about the analyzed samples.

Once the risk assessment had been carried out, the results need to be analyzed and interpreted. This leads to recommendations whether there is a need for risk managers to take appropriate measures. Depending on the country, the interpretation and use of the results of the risk assessment may differ depending on the criteria used. For example, Germany and the UK compare the use and the effects of GM

Table 4.3 Two different methodologies for detection of GM materials

Genetic modification usually involves insertion of a piece of DNA into the genome of the organism to be modified. Such foreign sequences can be detected both at the DNA or protein level. In both cases, quantitative and qualitative methods are available although with different sensitivities. Europe tends to use DNA detection methods and USA relies primarily on the identification of the expressed gene product, that is its protein

DNA-based method is primarily based on multiplying a specific DNA with the polymerase chain reaction (PCR) technique. Advantages of the method is that it quantifies molecules of interest in virtually all foods on the market today, quality of sample preparation is not important and it is a highly sensitive method. The disadvantage of the method is that it does not indicate whether the introduced DNA (gene) is active

Protein-based method is concerned with detecting proteins with antibodies usually against a specific protein, and enzyme assays that detect the activity of a protein. Advantages of the method is that it indicates whether the new gene is active and to what extent in the recipient organism, is quantitative and sensitive and does not need special training or new sophisticated laboratory equipment. Disadvantages of the method are that proteins are easy to degrade so that the quality of the extracted material is important. The method can also not be used efficiently on processed food

crops to conventional agriculture, while Austria or Sweden take organic farming as a source of comparison.

5.2   *What Happens If Risk Assessment Does Not Lead to Clear Answers?*

In many cases of potential environmental or health risks, the scientific knowledge base is not good enough to assess or interpret potential risks in a quantitative way and with sufficient certainty. For example, understanding of complex ecological systems may be lacking as is the knowledge to predict long-term effects on our health. This scientific uncertainty has profound effects on the way GM crops and their products are dealt with by risk managers and decision makers.

6   Risk Management

The role of risk management is to reduce or eliminate risks as identified during risk assessment. General risk management measures, also applicable to GM crops and their food products, could include the following:

* Preventive measures

    – Confinement strategies, for example, certain GM crops are only allowed to be grown in greenhouses
    – Re-evaluation of products, procedures
    – Restricted use, for example, the growth of GM crops could be restricted to certain geographical areas

- Delays in introduction onto the market (moratoriums)
- Removal from the market

- Establishing standards, for example, those concerned with drawing up of new guidelines or harmonization of approaches
- Monitoring following the commercial release of GM crops or their products
- Strengthened documentation requirements, such as those embedded in traceability of GM crops and food products
- Incentives
- Education, such as advise on good agricultural practice and technical support

However, in case of GM crops and their food products, scientific uncertainty assessing their potential risks and impact on human health and the environment created a challenge to risk managers and decision makers. To deal with this challenge, they have resorted to the use of a decision making tool called the precautionary principle.

6.1 Risk Management in the Face of Uncertainty: The Use of the Precautionary Principle

Very often scientific data is not available or is insufficient to assess a possible risk of a new GM crop. Some important questions may not be answered, for example, due to lack of data on fundamental biological processes. In other instances, scientific opinion may differ about the impact of findings. This insufficient knowledge or diverging views may be used to invoke the precautionary principle. It is a decision making tool available to risk managers to take appropriate (cautious) actions and measures when facing scientific uncertainties about a hazard that has as yet unclear likelihood and level and scope of impact.

However, precautionary decisions are based not only on the scientific information available, they are also based on socio-economic considerations. That is why precautionary principle is a political rather than scientific decision making tool. Because of inclusion of socio-economic considerations, precautionary principle is often discussed in relation to broader issues such as globalization and world trade.

Internationally, the precautionary principle was first used at the 1992 Rio Confe-rence on the Environment and Development in Article 15 of the Rio Declaration. "In order to protect the environment the precautionary approach shall be widely applied by States according to their capability. Where there are threats of serious or irreversible damage, lack of full scientific certainty shall not be pursued as a reason for postponing cost-effective measures to prevent environmental degradation." In other words, invocation of the precautionary principle is a proactive decision, on the part of the risk managers, to prevent possible damage to human health or to the environment. This includes ideally cost–benefit analysis of action or lack of action. The application of the precautionary principle should be nondiscriminatory and consistent with measures adopted under similar circumstances. It must be limited in duration because more new data is necessary to allow an objective final decision.

6.2 Political Dimension of the Precautionary Principle

Precautionary principle is not a scientific approach, it is a tool used by decision makers to make what at the end are active political decisions to use caution for the protection of the society and the environment in light of scientific uncertainty. The principle is thus meant to help facilitate management decisions. Decision makers do not only consider scientific evidence but, as already indicated, also social, cultural and economic needs of the region. Precautionary principle makes decision makers responsible for situations where previously no decisions were expected: under the precautionary principle they need to anticipate and prevent unacceptable harm to the society or the environment. In effect, the precautionary principle and the resulting precautionary actions amount to sensible actions on the part of public institutions, representing the society at large, to safeguard the public and the environment against possible harm. The precautionary principle is not an objective decision making instrument as it is a political tool biased to caution.

6.3 Why Should Precautionary Principle Be Applied to GM Crops and Their Products?

As genetic modification of plants has raised a number of health and environmental concerns, it is not surprising that there are attempts to apply the precautionary principle to GM crops and their products. The precautionary principle should be applied to GM crops, products, and processes for three main reasons:

- Because genetic modification of plants is a new technology with little baseline data available.
- Because genetic modifications can be inherited and propagated from generation to generation and thus potentially cause irreversible damage to the environment.
- Because GM food has, compared to our traditional food, not been on the market for very long and thus may pose some, as yet unknown, long-term risks.

6.4 When Should the Precautionary Principle Be Triggered?

Two prerequisites are necessary to trigger the precautionary principle. First, the identification of a potentially adverse effect originating from the GM crops or their products and secondly, the impossibility of assessing the risk with sufficient certainty, for instance, because of insufficiency of the data and/or their inconclusive nature and/or divergence of scientific opinion about the interpretation of the collected data. The implementation of the precautionary principle must be transparent and in line with existing chosen level of health and environmental standards of that particular country or region. The European Commission points out that precautionary regulations imply regulation of subject matter on the basis of standards that remain open for discussion. The regulation does not define these standards but sets a number of guiding principles.

6.5 What Happens When Precautionary Principle Is Invoked?

Precautionary principle, as used in Europe, has been used primarily to gradually phase in and re-evaluate GM crops and products in the name of health and environment protection. As it is applied on temporary basis, precautionary measures are reviewed in light of new information (Fig. 4.1). Thus new research is an integral part of precautionary action. When triggered, the precautionary principle also demands ongoing monitoring and risk assessment. Only on the basis of new information can recommendations be made whether or not to permanently ban, withdraw or allow a product, process, or a technology onto the market.

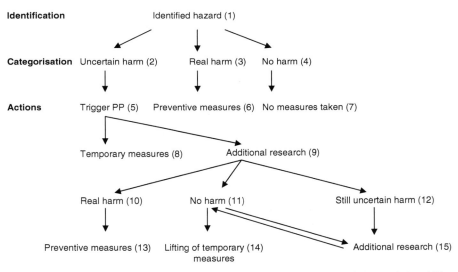

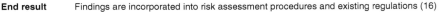

Fig. 4.1 Risk management: decision making process linked to GM crops (adapted from Pechan and deVries (2005))

Notes

(1) In order to trigger precautionary principle, the identified hazard (a source of danger) needs to relate to possible risks to human health and/or the environment. New hazards are usually identified based on new scientific results (discovery of the ozone hole is a good example). There are three possible actions to be taken, described in points 2–4. The newly identified hazard can relate to emerging or already existing technologies, products or processes that have already undergone risk assessment.

Fig. 4.1 (continued)

(2) Only in the case where there is uncertainty about the likelihood and/or impact of the risk associated with the hazard can the precautionary principle be invoked. Consensus needs to be reached on "triggers of action." The decision whether or not to trigger the precautionary principle should be also preceded by a wide ranging discussion by all the stakeholders.

(3) There is real and definite risk to our health and/or the environment, immediate action needs to be taken

(4) There is no risk posed by the threat, no action needs to be taken.

(5) Precautionary principle (PP) can be triggered only on the basis of uncertainty about scientific results. Uncertainty can have a number of causes: divergent scientific opinions (for example, about cause–effect relationships), insufficient data, wrong data, insufficient frame of reference (the type of questions asked and answered), etc.

(6) Preventive measures are usually entrenched in existing regulations. Application of these regulations has to comply with national and international laws and principles (see Table 4 .1).

(7) If, based on existing information, the conclusion is that no real risks are associated with the threat, no further action is needed and the proposed threat is re-classified as an observation.

(8) One of the two primary objectives of precautionary actions is to take temporary preventive measures to protect the health of the public and safety of the environment. There are a wide range of temporary measures that can be imposed, from outright banning to advisory statements. The choice of the temporary measures is made by the decision makers within the framework of national and international laws and principles (see Table 4.1). Although a political decision, it is be advisable not to take economic concerns into consideration when deciding on preventive measures, especially if to be used internationally.

(9) The second primary objective of precautionary actions is to carry out additional research to fill in the knowledge gaps. Its ultimate objective is to reduce the uncertainty about a possible risk to a level that allows a yes/no decision whether and how to act (see points 3 and 4). The greatest challenge is for all the stakeholders to agree on frames of reference (the questions to be asked and answered), the tools and methodologies for analysis and interpretation of the results. It is at this point that ethical and other considerations should become part of the research design and considerations. The fear of many is that either the frame of reference will be too broadly or too narrowly defined to sufficiently address the concerns about the identified possible threat. Additional research can have again three outcomes as defined in points 2, 3, and 4, namely:

(10) The possible threat is real and poses a risk

(11) There is no threat

(12) Insufficient data is available to make a yes/no conclusion

(13) In case of a real threat, a number of preventive measures can be undertaken, that are in line with national and international laws and principles (see also point 6).

(14) In case that the threat does not pose a risk to human and/or environmental safety, the temporary measures imposed by triggering of the precautionary principle are lifted and /or modified.

(15) In cases where no definite conclusions can be drawn, additional research needs to be carried out and temporary preventive measures likely remain in place. Many people are concerned about frames of reference that are too complicated, a priori open to subjective interpretations, that rely on additional socio-economic political considerations after additional research is concluded and that protract the time needed for reaching a definitive yes/no decision. This may stifle research, freeze innovation and reduce rather than improve transparency and effectiveness of the decision making process.

(16) The end result of triggering of the precautionary principle should be to incorporate the lessons learned into the mainstream risk assessment and management procedures. This will in effect add additional filters and considerations to the existing decision making procedures for identification of new potential hazards (see point 1).

6.6   Application of the Precautionary Principle
to GM Crops in Europe

GM crops are considered by many to be new and potentially harmful to human health and to the environment. Decision makers decided to proceed cautiously with the introduction of GM crops and their products. The end effect was the application of the precautionary principle in the form of a moratorium that was imposed for a number of years in the late 1990s. During that period of time, numerous studies were undertaken to assess (and re-assess) the safety of GM crops in Europe. This caution is reflected in the legislations governing the release and marketing of GM crops and their products. A number of precautionary legislations were implemented that relate to GM product safety both to the general public and the environment.

The European Directive concerning the deliberate release of GMOs (genetically modified organisms) into the environment can be viewed as the first legislation in which precautionary principle is translated into precautionary legislation. The legislature is structured to deal with each new GM crop on case-by-case basis using a step by step procedure.

The European traceability directive, also takes a precautionary view of GM products. It requires that all GM products can be traced from their source all the way to the end user. This is done in order to assure that, if something goes wrong, the product can be identified no matter where in the food chain and appropriate precautionary action taken. Traceability addresses issues of potential health risks and also risks of possible damage to the environment.

Related to the above is the labeling directive that requires labeling of GM products. It gives consumers choice in what they buy.

The European Union (EU) Novel Food Regulation 1139/98 ensures rigorous procedures and review processes are in place to ascertain that are safe for human consumption. To be nondiscriminatory, this approach applies to any novel products, whether or not they are of GM or non-GM origin. The invocation of the precautionary principle for existing GM products already on the market in the USA and those being imported into Europe could not be justified on health risk criteria as they have been shown to be safe with very low levels of uncertainty.

Socio-economic impact of the traceability directive

European Union traceability and segregation requirements are based on the precautionary principle as they foresee a possibility to withdraw a GM product from a market if a real harm is identified subsequent to its release. As it is structured, the legislation also addresses the risk of contamination of organic products by GM products.

The traceability requirements place responsibility on businesses to ensure that individual batches of GM products are individually traced from the farm all the way to the marketplace. The testing methods are now well established.

There are a number of arguments against and for traceability (and the associated labeling) legislation.

Main arguments against traceability of GM products

All products, including GM crops, allowed onto the market must be safe. So why trace and label them if they are already safe? Such actions may be misinterpreted as a warning and be prejudicial to GM crops. Although traceability may in theory reduce the need for continuous sampling, the reality shows that especially non-GM crops will need to be regularly checked for contamination from GM products. This will add costs to the food products. Moreover, developing countries may suffer the most from this type of legislation. For example, if the immediate need of a developing country is to reduce insect attacks on their crops, GM crops containing a Bt gene, may be one of the preferred options to address the problem. However, the developing countries may lack adequately trained personnel and testing laboratories, to comply with the product identity preservation and labeling requirements as outlined in the new EU directives. The difficulty with the traceability requirements may also be complicated from possible unintended contamination of non-GM crops or seeds with GM material. Rather than risk trading problems with the EU, developing countries may decide not to grow GM crops. Although developing countries should always compare alternative solutions to their problems, a priori exclusion of GM crops as one of the possible solutions, may deprive the country or region from effectively addressing, for example, specific insect or weed problems.

Main arguments for traceability of GM products

Traceability and segregation of products provide the public with a choice what to buy as it preserves the purity of the material. This is an important point for the organic industry. Finally, traceability legislation assures that if something would go wrong, the GM product in question can be rapidly identified and withdrawn from the market. So it is a prudent step to assure the highest level of safety.

6.7   Precautionary Principle Impact on Innovation and as a Tool for Reaching International Trade Agreements

Application of the precautionary principle and precautionary legislations have an impact on the international trade: first by impacting competitiveness of innovative companies and second by affecting international trade agreements.

Competitiveness of innovative companies

Precautionary principle asks decision makers to take actions in case of uncertainty. Uncertainty inherently asks people to be careful. Precautionary principle offers decision makers the secure option of "lets pause and wait for more data." There is no risk involved for decision makers to take such a view. Politically, such decisions may even be very popular, resulting in short-term gains and because the long-term harm of such a decision is not yet obvious. Indeed, precautionary actions to temporarily stop something are usually a better guarantee to protect their political careers that taking the "risk" to let something proceed.

By slowing down and complicating the approval process, new innovative technologies may fail to become competitive as this requires first and foremost speed, agility, and the right political climate. A way forward is to place the precautionary principle under the leadership of independent (apolitical) risk managers.

International trade agreements

In the context of international trade every country has a sovereign right to apply guidelines that it deems to be appropriate. Every country is autonomous in their decisions. However, all decisions have to be justified in the light of existing scientific data. In effect, a country needs to convince all others that its decision is fair and nondiscriminatory. This is true also in cases of precautionary actions where a management decisions are taken in view of identified potential risk to the health of its citizens or the environment.

The basic idea of precautionary principle is that it is triggered when a plausible threat to human health or environmental exists that is deemed unacceptable, without having to wait until the reality and seriousness of the identified risks become fully apparent. This assumes that there is a consensus on the meaning of risk and what we should do about it. Within a closely cooperating region, such as the EU, this is possible. It is much harder internationally.

The use of the precautionary principle on an international level faces a number of challenges that have primarily to do with drawing up appropriate frames of reference (criteria) that should in the first instance increase our understanding of the scope and likely impact of the identified possible hazards. Precautionary actions need to be carried out in order to be satisfied that GM crops and their products pose no health and environmental risks. However, the decision making bodies need first to agree on the following:

(a) What are the conditions to trigger precautionary actions (e.g., it needs to be agreed what constitutes a risk to human health and the environment and what level of protection is acceptable)?

(b) What are the key questions to be asked and answered (this will require additional research on case-by-case basis)?

(c) What are the methodologies, procedural rules, and standards to be used in the investigations (so that at the end we can say: now we know enough, with sufficient scientific certainty, to make a permanent rather than a temporary decision)?

Disagreements between countries can arise due to different definition and interpretations of these frames of reference. For instance, definitions of what constitutes an acceptable risk are very much country and region specific and dependant on local socio-economic conditions, moral values, and collective experiences.

In its present form, the precautionary principle does not attempt to reconcile nor address these differences. This seriously limits its international appeal.

Moreover, by incorporating socio-economic considerations into the precautionary principle and by placing the principle into the political realm, vested national or regional interests may be allowed to define the frames of reference to suit their needs.

Different countries may have different criteria of what level of safety is acceptable in light of a possible harm arising from an identified hazard. For example, societal values and attitudes influence these frames of reference, not least definitions of hazard, threat, risk and harm and standards. This is well illustrated by the difference in attitudes between the EU and USA, vis a vie the precautionary action and precautionary principle.

Both the precautionary principle and precautionary approach appear to follow the general principles of risk management that includes proportionality, consistency, and nondiscrimination. However, precautionary principle and precautionary approach may create contradictory needs, objectives, and interpretations of results (see Table 4.4). This can ultimately lead to international trade conflicts.

The choice of one word can have far-reaching political consequences. This can be seen in the difference between the words precautionary *principle* and precautionary *approach*. The USA sees no need to acknowledge the EU precautionary principle as an international standard because the USA claims precautionary approach is already inherent in all science-based analysis to assess risk.

Although both recommend precaution in dealing with GM crops and products, the actions based on European Union precautionary principle differs from the actions permitted in the USA using precautionary approach. The frame of reference and philosophy between the two concepts differs significantly.

Table 4.4 Precautionary principle vs. precautionary approach: Two main differences between precautionary actions in EU and USA

EU precautionary principle	USA precautionary approach
Considers societal value systems in addition to scientific evidence	Is said to be based on scientific evidence only
Takes health and environmental sustainability issues into account	Health and environmental safety is not discussed in terms of long-term sustainability

One possible scenario under the precautionary principle is that if socio-economic aspects take precedence over scientific reasoning, USA or Europe could justify their opposing positions on the basis of regional needs. The end result is that products allowed on the market according to one set of standards, may be rejected on the basis of another. Emphasis on ethical rather than socio-economic aspects may place the deliberations on an equal playing field.

A possible way forward is to allow Codex Alimentarius, a joint body set up by the Food and Agriculture Organization (FAO) and the World Health Organization (WHO), to conclude new (risk related) definitions and standards for GMOs. As the standards agreed upon by the Codex Alimentarius are usually referred to by the WTO, this may open a way to reach binding international agreements. This process is, however, very time consuming. A second instrument is the UNEP Cartagena protocol on biosafety to the Convention on Biological Diversity. It emphasizes the need to accept human and environmental concerns first: to accept that they take precedence over economic considerations. Without an international agreement on the meaning and application of the precautionary principle, there is a real possibility that trade wars may erupt in the coming years.

In this respect it is interesting to note that Codex Alimentarius Committee on Food Labeling (CCFL) is discussing process-based labeling of GM foods. There appears to be growing consensus that labeling is needed for foods derived from modern technology when there are significant changes in composition, nutritional value, or intended use and that it is important to provide such information to consumers. The CCFL has also achieved a consensus on the labeling of allergens in foods derived from modern biotechnology and believes that such provisions provide considerable assistance to and protection for consumers.

6.8 Concluding Comments

The precautionary principle should be triggered only in instances where identified concerns can be measured. Concerns need to be prioritized and must relate first to health and environmental risk, with science and ethical considerations at its core. The challenge is to come to an agreement of what constitutes a hazard and risk as these are the triggers for precautionary actions. Although science is not always objective and results are subject to interpretations (indeed that is one reason to invoke the precautionary principle), science is nevertheless one of the best tools we have to reach fair and transparent decisions that are respected across cultures.

It is important that there is input from scientists in drawing up the "triggers of action." This should be based on a combination of state of current knowledge, risk, and benefit comparisons to other comparable existing technologies, products, or activities in cohort with a set of ethical criteria.

The frame of reference for imposing precautionary actions should be defined according to the present status quo and compare possible risks and benefits of a new technology or product with those of available alternatives, that in turn should also be subject to risk/benefit analysis.

The primary aim of the precautionary principle should aim to reduce uncertainty in scientific results. Risk management decisions must be apolitical, carried out by independent risk managers and devoid as much as possible from socio-economic considerations. This to assure transparency, fairness and impartiality in the decision making process.

There is a need for urgent international action on a number of issues put into focus by the precautionary principle debate. Among them are agreements about various risk-related definitions (triggers of action), risk assessment procedures, standards, competencies, possible misuse of the principle and its oversight.

If universally accepted, the newly formulated anticipatory actions may even be applied more broadly and could even be used in political decisions of managing world peace and international trade.

7 Communicating GMO Issues

Decision making at all levels, personal or societal, must always weigh risks and benefits of a decision. Today, many of the technologies we use are complex. Increased knowledge of how these technologies work often raises uncertainties, especially when the technologies have multiple applications. Risk decisions, especially when based on uncertainties, increase the level of possible risk. Precautionary actions may be called for as well as an in depth risk assessment of the risks in order to eliminate or reduce the uncertainty about the risk effects. Constructive dialogue between the key stakeholders is essential to become aware and familiar with the concepts of risk and uncertainty. Modern biotechnology is a highly sensitive topic, both socially and politically. Today, public opinion is not merely a perspective, it is a constraint on the ability of governments and industries to exploit new technologies. It is thus important to openly discuss genetic engineering of crops, where real and perceived risks may influence our decisions.

There are two important aspects to consider when communicating GM crop issues: first, the content of the message, and second, the strategies of presenting the information to the public. Both influence what the public may think about a certain topic.

Information to be communicated should be accurate, complete, easy to access, and understandable. Communicators often do not empower the public to make knowledge-based decisions. Rather, they may pursue specific interests. This means that the information brought to the public is sometimes selective and not objective.

7.1 The Challenge of Communicating Risk Issues

As already eluded to in Chap. 1, our society is faced with an interplay of physical, environmental, economic, and social risks. Our society is becoming ever more complex and, in some respects, more vulnerable. There are fewer companies controlling more of the market – this is primarily because of the ever-increasing integration and globalization of society. If something goes wrong, because there are fewer players in any one field, the impacts can be catastrophic for the whole society. The OECD refers to these risks as systemic risks. Systemic risks are at the crossroads between natural events (partially altered and amplified by human action such as the emission of greenhouse gases), economic, social, and technological developments and policy-driven actions, both at the domestic and the international level. This makes dialogue about risks difficult – both the experts and the public are often unsure of the trans-boundary effects of these risks. The public is uneasy about these developing patterns.

For example, various areas of biotechnology, such as GMOs, are seen as highly controversial, not only because of moral concerns, but also because of who controls the GM seeds and how the GM products are marketed around the world. Thus the dialogue about GMO risks also touches on issues of free trade, the power of industry, globalization, sustainability, monopolies, and rights of individuals. Moreover, decision makers, including risk managers in Europe and elsewhere in the world, are generally not trusted to be in control if something goes wrong with GM crops or their products.

Risk communication about biotechnology issues is complicated by the fact that the same technology may be perceived differently depending on its application. Genetic engineering applied to human diseases is, with a few notable exceptions, generally welcomed by the public, yet the benefits of genetic engineering of plants is not as obvious to the public and has been generally opposed on grounds of potential environmental risks.

7.2   Attitudes to Risks Concerning GM Crops and Products

The perception of and attitudes to risk differs between risk experts and the general public. While risk experts base their risk decisions on assessment of available scientific knowledge, the general public generally decides about risks on the basis of their attitudes to risk. Overall, attitudes toward risk are a sum of three components: Cognitive (knowledge), affective (emotions) and conative (related to actions and activities). Below are given key examples of the three components as they relate to GM crops and products.

7.2.1   Conative Element

An important aspect in helping to form attitudes toward GM products is freedom of choice. Generally, the less freedom of choice, the higher the perceived risks. That is why the G labeling of GM products provides a choice to the consumers and reduces the feeling of danger or imminent harm. One can always "escape" and buy the non-GM product. This is an example of a conative (action based) component that helps to form our attitudes.

7.2.2   Affective Element

General public is more fearful of unknown risks. This is especially true if the potential risks are prolonged or perhaps irreversible (e.g., GM crop effects on ecosystems). The more risks are understood, the less the public fears them. The explanation of potential risks should be done by communicators that the public trusts.

Genetics and biotechnology are not easy to understand. The public needs to trust the communicators explaining the pro- and contrapoints of GM crops and products. As already eluded to above, politicians cannot be trusted with this job. There is a general mistrust in the decision making and supervisory bodies as well as large companies. The trust in scientists is still high. Trust is dependent on: competence of those communicating the issues, objectivity, balance, completeness of argument, impartiality, fairness, consistency, and belief in the communicators integrity. Indeed, these are values that define successful science journalists and science communicators. A source that is trusted and believed in can be persuasive enough to achieve changes in attitudes.

Another example of the affective component are human values. Value orientation is expressed in political preferences, lifestyles (respect of nature), awareness of environmental issues and religious convictions. They influence for example how media information is interpreted (perceived) in regards to GM crops. This mechanism may selectively filter and prioritize what information is made available for knowledge-based (cognitive) evaluations of risks associated with GMOs. Combined with the human tendency to pay attention to negative (and higher risk) news, communicators with positive news about GM crops may face an uphill battle to change the pre-existing negative attitudes.

7.2.3   Cognitive Element

Knowledge (cognition) is the third element that helps to form attitudes. Availability of relevant information and previous knowledge motivate a person to take time to make an in depth analysis of the situation (pro- and contra-arguments) and make an appropriate decision, for example, whether GM maize is safe for the environment. When lacking such information and knowledge, a person will likely make a superficial decision that is more based on previous personal experiences (e.g., negative experiences with large companies leading to mistrust) rather than analysis of the current situation based on available evidence. Attitudes and subsequent actions may thus be influenced by the motivation of the person that is in turn influenced by the availability of relevant information and accumulated knowledge. In general, the attitudes of the general public are less formed by the careful analysis of concrete risk assessments, rather they are formed by general mistrust of the possible misuse of the technology in the hands of large companies.

In conclusion, an individual attitude to GM crops and their products is dependant on a number of subjective assessments, influenced primarily by their personal experiences and values. A change in attitudes in the majority of Europeans will be required in order to achieve market acceptance of GM products derived from GM crops. This will need to include changes in knowledge, emotions, and actions. Of the three components, the emotive issues seem to be the hardest for proponents of GM crops to successfully address. It is thus not surprising that the discussions about GMOs, even after more than 20 years, do not seem to lose their intensity. Had only facts played a role in forming attitudes to GMOs, the discussions would have been less emotional and likely no longer of public interest. Three key characteristics of GM crops seem to keep the discussions in the public forum. One is mistrust in large companies that are not accountable to the public, yet have the power to irreversibly (and negatively) affect the environment. Second is the limited usefulness of GM crops for the end consumer in the developed world. Third is to provide consumers with choice between organic, conventional and GM food products. If proponents of GM crops are to change the current negative attitudes toward GM crops, they need to address these three issues more successfully.

7.3   Why Communicate GMO Issues

There are a number of reasons why communicating GM crop issues this needs to be done. The obvious reason is that without a constructive dialogue, GM crops and products will have a hard time to establish themselves in many parts of the world. It maybe that in many industrialized countries, GM crops are not needed, simply because there is already agricultural overproduction. But the situation maybe different for other parts of the world where food is scarce or expensive. Provision of healthy living conditions may mean to use GM crops in areas of the world where climate change is putting conventional agriculture under pressure. Lack of water resources, for example, calls for crops that can withstand drought and high salinity conditions. To help feed the world is not an option, it is a moral obligation. However, any decision to improve persons well being is linked with weighing risks and benefits. These include at the decision making level analysis of risks of using various forms of agricultural practices (organic, conventional, and GM) to feed the hungry. These issues need to be discussed with the public.

The complexities of risk deliberations, the various definitions and dimensions make these discussions difficult. Nevertheless, constructive dialogue with the key stakeholders is needed to minimize unfounded concerns and maximize public empowerment on these complex issues. GMOs are a current example of new technologies that are penetrating our daily lives. In the future, there will be other technologies and issues that may impact our daily lives, and just as complex. Thus the dialogue about risk issues stands at the very center of a democratic society. The deliberations reach all of us, both as individuals and as a society.

7.4   Who Should Do the Communicating and Where Should It Take Place?

Originally, GMO issues were communicated top-down, for instance from a regulator or industry to the public. More recently, a dialogue form of risk communication which encourages public and stakeholders to actively participate in the communication process has become preferred.

The decision of who should communicate depends on the target groups. Indeed, the choice of the target group will also dictate the content of what should be communicated and how it should be communicated. In case of GM crops and their products, the main stakeholders are scientists, industry, NGOs, general public, and decision makers. If the target group is the general public, the method of presentation and content will differ than if the target group are school children. In the former case, it may be important to solicit the services of a high profile public figure to communicate the key messages. In the latter case, teachers would be the communicators. Indeed, teachers can also be the target group as they would need to be brought up to date on key issues and topics.

More sophisticated communication strategies subdivide groups within the general public, for example, on the basis of their age, sex, interests, education, income, language, or attitudes. The category of those that need to know more also includes decision makers and others who through their profession need to be informed about risk issues.

Mass media, schools, and professional organizations are the likely platforms for communicating risk issues. NGOs rely on the use of mass media: predominantly printed press, television, or radio. Judging from public surveys the most effective means to reach the public is through television. This knowledge has not been lost on NGOs: they have learned to effectively gain the spotlight in television news by carrying out very specific and visually grasping actions.

7.5   What Should Be Communicated

One key point to remember in communicating GMO issues is that the general public is more concerned about involuntary than voluntary risks and more about technological than natural hazards. GMOs, in the food context, fall under involuntary and technological risks. This combination is a real challenge for the communicator. The communication has to include discussion of potential impacts on consumers.

The choice of message content presentation, in a way that the public not only notices but positively reacts to the message, relies on the understanding of human psychology. NGOs have learned, as advertising experts and film makers have known for a long time, that a primary way to reach an audience is through their emotions and to play primarily on their fears. Some NGOs that want to stop the development and marketing of GMOs in general, and GM crops in particular, label GMOs as unnatural and in the hands of monopolies. It had taken on the clothes of Frankenstein and GM crops became "Frankensteins food." The story of Frankenstein represents and invoked memories and feelings of pity and fear and anger. Fear of Frankenstein (but also pity for him) and anger for the folly of his creator. In case of "Frankenstein food" the counter transference often relates to the viewers childhood fears of the unknown, of darkness, of death.

To bring across a positive message, that GM crops and their products may be useful and safe, is much more difficult. This is especially the case as to date there are no clear examples of really useful GM products for the general public in industrialized countries. While all effort is made to minimize hazards occurring, food safety is not an absolute and hazards can occur. Thus introducing something onto the market is inherently more difficult than trying to keep it away. Unless one can successfully argue that keeping something away from the market place is more dangerous for the public. What is clear is that better public relationship strategies are needed to restoring the trust of the public in the technology. First and foremost this means to be able to convince the public of the urgency and importance of the chosen topics and secondly to dedicate sufficient time (likely on television) to discuss these issues in a way that will sustain the public interest.

8 Addressing Potential Risks of GM Crops: Case Study of Germany

The struggle over the use of plant genetic engineering has been carried out at different levels of the society. Not only can this be seen in open public debates across Europe, the discussions were also transported into the media and politics both at the local and national level. The outcome of these discussions, in many instances real power struggles, resulted in a number of decisions that impacted the use of GM crops in Europe. An example of how GMOs issues were dealt with in practice in one European country – Germany – is given below.

For clarity, this overview is divided into three sections, namely:

1. Public opinion and media
2. Politics
3. Regulations

It spans the years from mid-1990s until late 2008. In this section, reference will be made to GMOs rather than GM crops or GM products. This is because in Germany, GM crop related issues are discussed primarily under the heading of GMOs.

8.1 Public Opinion and Media

German attitudes to biotechnology, just like in other countries, are shaped to a large extent by the area of its application. Thus while the applications in medicine are generally seen as positive, applications where no immediate benefit to the consumer could be shown were seen more critically. The "green" biotechnology falls into this category. Germany belongs to the countries that are especially critical of GMOs. Majority of the population rejected and still rejects green biotechnology.

In mid-1990s two developments influenced the public perception of GMOs in Europe. The first was the large-scale planting of GM soy in North and South America (and its subsequent marketing around the world). The second was the cloning of the sheep "Dolly" in the UK. These developments had a significant influence on media reporting and public discussions. Both were seen as a potentially negative development. The end result was that by 1997 the public acceptance of biotechnology had reached a low point. This despite the fact that the press reports about GMOs in Germany tended to be neutral and well reasoned. The negative attitudes to GMOs in Germany were fuelled by negative comments primarily by Greenpeace and the political Green Party – about possible health problems, irreversible damage to the environment or loss of food quality through serious problems with "contamination" of our food supply with GM-modified products. Indeed, the German Green Party that became a junior partner in the government in 1998 sometimes seemed to follow the lead of Greenpeace. (In Germany, Greenpeace has more members than the Green Party). The effect was that the debate about GMOs became politicized. Risk issues became only part of the equation of determining attitudes to GMOs. At the end of 1990s green biotechnology ended up on the loosing end of public popularity in Germany. About 75% of the population rejected green biotechnology. Indeed, even Europe-wide, only 32% of the citizens though that green biotechnology would improve their lives.

The start of the twenty-first century had made the situation for GMOs in Germany even more difficult. This had to do with a number of badly handled public health issues, for example, the BSA outbreak in the UK. The politicians, some scientists, and industry had initially played down the scope and dangers of the outbreak. This had further eroded the trust of citizens in the decision making process in Europe, posing the question: "if they can't get this straight, what could we expect with GMOs?" As in previous years, around 75% of the German population rejected GMOs.

Nevertheless, the start of the twenty-first century turned out to be a turning point for biotechnology in general. In 2002, Europeans started to see the benefits of biotechnology more positive. More than 40% of Europeans thought something useful may come out of the technology. That was 10% more than in 1999 and even more than in 1996 (for details see the appropriate Eurobarometer reports). There were nevertheless significant differences between countries and social groups in the acceptance of GMOs. Men saw biotechnology more positive than women and younger more positive than elderly. Countries such as Czech Republic and Spain saw more benefits than risk sin the use of biotechnology, while the population in Germany and UK was much more skeptical.

The more recent analysis of attitudes to GMOs, carried out in 2006, showed that nearly 75% of the German population still rejected the development and introduction of GMOs onto the market. Only 7% approved of this technology and 18% are neutral. The application of biotechnology outside of the "green" area was viewed much more positive.

At the heart of the problem with green biotechnology in Germany is that the public has not really seen any need for it: there are no obvious benefits over and above conventionally or organically grown crops. Moreover, increasing productivity of crops is not seen by the general public as something of importance. Europe and North America are used to food overproduction and agricultural subsidies. Europe, including Germany, has already enough food without the introduction of GM crops. Proponents of green biotechnology were simply not able to find the right arguments to convince the general public of the usefulness of this technology. However, this situation may change in view of the rising oil and commodity prices.

On the other hand, in case of organically grown crops in Germany, good marketing strategy, and political support have succeeded in portraying organic farming as something good for your health and good for nature. This strategy was combined with adjectives such as traditional and natural. Organic farming claims to respect nature, support local producers, and thus be ethically correct. The overall effect was that a contrast was established between organic and GM food products, the former representing something wholesome, healthy and correct, the latter as something possibly dangerous, big business oriented, and ethically questionable.

The claims for organic farming were made without really questioning or examining the underlying assertions. It is only now that scientists both in Germany and elsewhere in Europe have been getting sufficient funding to really compare organic, conventional, and GM grown foods for their composition, nutritional values and safety.

One unique point about attitudes toward biotechnology in Germany is the angst of its misuse. This could be linked to the experiences of Germany during the Second World War where misuse of knowledge was institutionalized and where objectives of the government justified destruction of human lives. The fear of things going out of control, even in a civilized society, has had a profound effect on German attitudes to societal safety issues. This attitude is not applied just to biotechnology, but also to issues such as intrusion of government into private lives of their citizens and safety of personal data and nuclear energy.

8.2 Politics

The political climate in the mid-1990s, in as far as the GMOs were concerned, was very mixed. The parties in power, Christian Democratic Union (CDU) and the Liberals were generally for bringing GM products to the consumers. For example, Helmut Kohl, the German Chancellor at that time, had come out in support of GMOs. At the opening of the "Anuga" food fair he, for instance, pointed out that GMOs are economically important for Germany, and that he was in support of their application in agriculture. The governmental opposition, led by the Social Democratic Party (SPD), requested labeling of GM products. The Green Party, as already mentioned above, had rejected plant biotechnology and had emphasized organic farming instead.

After the federal elections in the year 1998 the government changed. Social Democratic Party (SDP) together with the Green Party came into power. This had a profound effect on how GMOs were dealt with. Although the Social Democrats were not in principle opposed to GMOs, the Green Party was vehemently against the GMOs. This led to diverging strategies how to deal with the GMO issues. GMOs became a political rather than a risk-related issue.

The then German Chancellor, Gerhard Schröder, a Social Democrat, proposed in the summer of 2000 a 3-year program for the evaluation and implementation program for the use of GMO in agriculture. This program, named "Kanzler Initiative zur Grünen Gentechik," was to be carried out in collaboration with industry. The idea behind the GMO program was for industry to limit itself to research and evaluation activities. In the meantime, true to the German tradition, a new consensus should have been found how to proceed with the commercial application of GMOs. Initially, industry should have had talks with the federal Chancellery and, after this consultation process, other key stakeholders would have been brought into the process. However, until the fall 2000 no discussions were held. The industry decided for itself that they would like to go ahead alone with the planting of GM maize without the political consultative process. In 2001, the government withdrew this initiative. Part of the justification for this action were also increased consumer fears as the result of the BSE crisis. Moreover, new agriculture initiatives made it nearly meaningless to start the program.

Ministries of agriculture and environment went to the governmental junior partners – the Green Party. Their political agenda was structured to make sure GMOs were not planted in Germany and if possible, do not reach the German consumers. Their strategy was implemented at different levels. Changing key personnel responsible for the testing and release of GMOs, changing the responsibilities of organs responsible for GMOs and creating new regulations that on one hand supported organic farming and on the other hand made it difficult to plant GMOs.

For example, in 2004, the Green Party environmental minister at the time, Jürgen Trittin, said that it is correct that a lot is being said and written about the release of GMOs into the environment. He saw as problematic the coexistence of GMOs and organic products. In essence, it was up to the farmers that wish to plant GMOs to make sure that they do not contaminate organically grown crops. Moreover, he stressed that we need to much more take into consideration the dangers that GMOs pose to the environment.

After the new elections in 2006, the CDU came back into power, this time in a co-sharing arrangement with SPD. In the coalition agreement, it was stated that commercial planting of GMOs should be allowed. This position was especially supported by the new German Chancellor Angela Merkel. The CDU reinstated their clear and positive position about the benefits of GMOs for the German agriculture. To assure free choice, in their opinion, the coexistence of GMOs would need to be possible alongside conventional and organic farming. The Minister of Environment, Sigmar Gabriel requested that the freedom of choice of the consumers and the possibility of coexistence must be possible. A number of CDU members of parliament were of the opinion to change the laws governing GMOs to make it easier for industry to do business with these products.

In 2007, the German Chancellor, Angela Merkel spoke out in support of the necessary changes. In August 2007, the Federal cabinet approved changes to the law governing GMOs as proposed by the Minister of Agriculture. However, later political developments have again put a stop to GMO use in German agriculture (see Sect. 8.4).

8.3   Regulations

There were and are many ways to influence the debate and destiny of GMOs in Germany:

- Change the laws
- Change the institutes responsible for GMO releases
- Change the personnel responsible for experimental releases
- Change the rules for GMO release applications
- Sue those that plant GMOs
- Go on media offensive
- Destroy experimental plots with GM crops

In this section, only the first issue will be dealt with – the politics of changing the laws governing GMOs in Germany. It suffices to say that all governments when they are passionate about a certain issue will not shy from using the first four means of influencing the outcome of debates. The last three listed ways are extreme, but they had been used in Germany to intimidate those who wished to plant experimental plots with GM crops. It goes without saying that without experimental releases of GM crops, not enough knowledge would have been created to prove that GM crops are safe to use (as demanded by the opponents of GMOs).

EU regulations, agreed upon in Brussels and Strasbourg, supersede national regulations. EU regulations become enforced automatically at national levels. Nevertheless, national laws and regulations do exist. This is also the case with GMOs.

One of the first laws that dealt with GMOs in Germany was enacted in 1990. It dealt with the protection of living organisms and the environment against the negative effects of GMOs. It included provisions for preventive actions against these risks. It also included provisions for the research and development GMOs products.

This law was modified in 1993 subsequent to the implementation of the EU regulation on the release of GMOs into the environment (90/220/EC). The procedures to test new GMOs were simplified, participation of the public reduced and the guidelines for the institution in charge of the GMO issues relaxed. It took nearly 10 years (2002) for this law to be modified again. These included only minor changes.

In the meantime, the 90/220/EC regulation was replaced by a new regulation titled 2001/18/EC. Its primary objective was to make the procedure for the release of GMOs into the environment more transparent and effective. The permissions for the release of GMOs into the environment would be given for 10 years while integrating monitoring procedures. The precautionary principle was taken as the guiding principle for the protection of people and the environment.

In 2003, a new EU regulation was published (Nr1879/2003) that dealt with the use of genetically modified food (for use by consumers) and feed (for use by farm animals). It set out procedures and regulations for approval and labeling of GMOs.

Partly in response to these two regulations and partly as a result of the long-standing efforts of the Green Party, a new law governing the use of GMOs was passed in Germany in 2004. A central point of this law was the protection of genetically unmodified conventional and organic agricultural products against "contamination" by GMOs at all stages of the agricultural production process. Under this law it was the responsibility of those who wish to release or sell GMOs to make sure that human health, environment, and purity of non-GM products is not compromised. All intended GM releases must be registered at the place of the release. Moreover, those who wish to release GMOs into the environment are legally liable for damage to surrounding farms if the GMOs "contaminate" genetically unmodified products.

In February 2007, after new German elections in which the Christian Democratic Union won, preparations got under way to change the law governing GMOs in order to create a more "fair balance of interests." The objectives were to

- Strengthen the research into GMOs
- Streamline the procedure for the release of GMOs
- Define good practice procedures
- Create transparency
- Clarify liability regulations
- Secure environmental protection

In August 2007, the cabinet agreed to the proposed changes to the GMO law. In 2008, a new law governing GMOs was passed. The main features of the law are as follows:

- Food producers can label their commodity as non-GM even if GMOs or their products were used during the preparation of the commodity.
- Animal-based foods can be labeled as non-GM when the animals were not fed any GM feed such as GM soybeans. Animal-based foods can also be labeled as GM free even when GMOs or their products are used in the preparation of the commodity. For example, GM enzymes can be used but they must not be detectable in the final commodity. This is allowable when no other non-GM alternatives are available.
- Setting certain coexistence limits in agriculture. For example, the distance between GM maize and organically grown maize must be minimum 300 m. The distance between GM maize and conventionally grown maize must be minimum

150 m. As in the past, the GM farmer is legally and financially responsible not to "contaminate" organic and/or conventionally grown crops.

The German consumer associations welcomed this law as it brings the consumer the choice to choose between GM and non-GM products. The Green Party was not happy with the law.

Finally, it should be noted that the EU regulation 2001/18/EC should have been adapted and incorporated into the German national regulations already in 2002. However, in the summer of 2008, only part of this regulation has been implemented. This is still the after effect of the past struggle between the Red-Green government of 1998–2004 and the Bundesrat (German senate) dominated by the Christian Democratic Union that did not wish to agree to the governmental GMO proposals.

8.4 Risk Assessment, Politics, and Decision Making

Under the Red-Green government, scientific advice played little part in GMO risk governance. These decisions were politically driven, especially as the Green Party, that still remains vehemently opposed to GM products, was in charge of two sectors closely linked to GMOs – agriculture and environment. The one decision making tool often used at the European level to deal with uncertain situations, namely the precautionary principle, was only selectively used at the national level. If used, it was used to support a specific political and ideological perspective on the GMO issue. Thus the idea entailed in the precautionary principle, namely to be on the side of caution, was frequently used to further Green Party perspectives on the GMOs and to influence the public opinion. However, the other key aspects of the precautionary principle, namely objectivity, balance, the need for further research and nondiscriminatory actions were generally ignored. Organic farming was pushed to the foreground while at the same time field research on GMOs was suppressed. This was evident not only through the changes made at the federal level in the personnel and decision making structures and processes related to GMOs (ultimately slowing down for some time gathering of information on GM crops from field experiments) but also in some cases destruction of GM crop plots and lawsuits by anti-GMO organizations to block planting of GM crops.

The decisions about GMOs were not based on the principle of risk governance, rather they were made primarily on the basis of political ideology. The ultimate question was not whether GMOs were safe or not, rather, the underlying discussion was whether we want to live in a society where GMOs could become part of our lives. Even within the ruling parties in Germany there are regional differences as to the support for GMOs (see Table 4.5).

In conclusion, although at the EU level well-balanced legislations were passed, they were not always fully implemented at the German national level. Here, different political interests led to postponements and interruptions in the testing and marketing of GM products.

Table 4.5 The politics of GMOs

In 2005, as federal minister of agriculture, Horst Seehofer supported the release of GM maize in Germany. However, once Mr. Seehofer became the leader of the Christian Social Union (CSU) in province of Bavaria (a sister party to the larger Christian Democratic Union), he started to oppose the release of GM crops. This opinion change was driven by the fears of the CSU about the steady lose of votes in Bavaria. As most Bavarians oppose the release of GM crops, the Mr. Seehofer 180° opinion change could be seen as to appease the attitudes of the voters. Basically, the Bavarian CSU was trying to win some voters back. Indeed, while trying to regain foothold in some of the farming communities, the CSU Environmental Minister of Bavaria went as far as saying that Bavaria should be a GMO-free zone. The Bavarian CSU position on GMOs is in direct conflict with the official line of CSU at the German federal level. Such discrepancies undermine trust of the general public in politicians and their decisions. In the specific case of GMOs, decisions based on risk assessment take a back seat to populist and opportunistic regional and local political decisions that ignore German federal governmental policies and EU directives. This ultimately negatively affected the federal CDU-CSU policy on GMOs.

Risk assessment is part of the precautionary approach to evaluate new technologies. Clearly, as the situation with GMOs shows, application of the precautionary principle loses its value as a decision making tool in politically charged situations. Instead of decisions based on scientific evidence and risk evaluations, what is often left are political statements that misuse the precautionary principle for reflecting political purposes.

9   Future Prospects

An ever-increasing number of countries grow GM crops. Land under GM crop cultivation has passed 100 million hectares. What will the future bring? GM crops are not going to go away.

GMO-related risk issues can be discussed from many perspectives. Arguments can be put forward to either support or reject GM crop and their products. To date, it has not been possible to convince many people of the usefulness of GM crops and products. This is primarily due to the fact that the benefits seem to benefit a rather limited group of users, namely the industry and the farmers. Future should see GM products that will likely be seen as beneficial by the population at large. The argumentation for their application is gaining momentum. More functional foods are being created. GM crops could have a positive impact in countries most affected by the oil crisis and the climate change with associated problems in marginal lands. Those arguing against GM crops point to the possible health and environmental risks. It is the fear of their far-reaching and all-penetrating impact in our every day life that makes many people weary of applying genetic engineering to agricultural products. GMOs are one of the best examples of a debate about technology transcending risks. In such complex situations, the public needs to trust their decision makers. Yet the public in many parts of the world has little trust in governments and big business. They are often seen as too powerful and not responsive of citizens wishes or concerns. On the opposite end of the spectrum, the high trust of citizens in North European countries in their governments is also reflected in their confident manner of dealing with GMO issues.

For many, especially in industrialized countries, the debate is about much more than just the balance of risks and benefits. It is about lifestyle choices. This can be seen in many European countries, such as Germany. People ask: what do we want our society to be like? For many, this seems to be a choice based on their lifelong experiences and moral views. The situation is often presented to the public as the choice between on the one hand organic "natural" foods produced locally by small farmers and, on the other hand, GM food products of uncertain safety produced by big business and shipped in from all corners of the world. The public bring the discussion down to their local level and, intuitively, they stand on the side of the weak and threatened and the "natural." It brings to the forefront argumentations that are more emotive rather than science based. Moreover, as the knowledge about the food we eat expands, it is likely that what we today consider as facts will need to keep pace with new discoveries and be updated. Knowledge is evolving. This may, however, pose problems to some members of the general public who seek to be fully reassured (the human desire for certainty is a well-known psychological characteristic). The changing nature of knowledge may thus increase the perception that GM crops and their products are unsafe and that experts do not really know what they are doing. This could ultimately increase, rather than decease, the insecurity and mistrust among the population in new technologies. Ultimately, in a knowledge-based society, the application of new technologies should be determined by evidence weighing the benefits and risks for a particular user group, while taking into account impact on the whole society and the environment. This is in contrast to populist decision that may encourage legally questionable actions.

10   Chapter Summary

Judging GM crops strictly from a food safety perspective, GM crops currently on the market are safe. But there are more factors that determine whether a new technology is accepted by the society. As specific uses of biotechnology illustrate, the public reaction to new technologies is governed by at least three factors: Whether or not the specific application is useful to the consumer and whether there are any safety concerns coupled with the trust in the regulatory institutions to manage the risks satisfactorily. In considering all these three factors, genetically modified organisms are not seen by the general public in a very positive light, at least in many industrialized countries. Indeed, usefulness and safety as well as trust in decision makers, are prerequisites for the general public not just to accept GMOs but any new technology, invention process or emerging risk.

Dialogue about uncertainty, risk assessment, and management is crucial to help the general public become aware how risk decisions are made and what role science and politics plays in these decisions. When do we act and when do we not act on a possible emerging risk? What are the costs of acting and not acting? How are these decisions made? How objective are scientific opinions – how much are they influenced by market pressures? What role does ethics and morality play in risk decisions? What are the drivers that help to form public opinions about new technologies? How much are politicians engaged in populism rather than public engagement? What is the meaning and role of democracy in the new age of scientific discovery? These questions go well beyond the specific issues surrounding the safety of GM crops, beyond our environment and health concerns. But what is clear is that the public needs to be actively engaged in debating such issues. The resulting answers and conclusions will shape all of our futures.

Glossary

Antibody: Monoclonal and polyclonal antibodies can be used for specific detection of the product on the basis of classical (immunological) antibody–antigen reaction.

Agrobacterium: Is a gram-negative bacteria used in the transfer of DNA between plants.

Bioethics: Addresses and resolves possible conflicts between factual information and morality. It examines crucial issues both in terms of appropriateness of choices and actions. It is a subject where science, philosophy, and law meet and deals the conditions and constraints under which we should apply new biotechnologies.

Biotechnology: Application of biology, including the field of genetic engineering, to our everyday lives.

Certified reference material: Are measurement standards for testing and analysis of materials to ensure reliability and comparability of measurements in these field.

Criteria of risk: Frequency and scope of a harmful event occurring.

Cocultivation: Cultivation of two types of cells in the same medium.

Codex Alimentarius Commission: The Commission was created in 1963 by FAO and WHO to develop food standards, guidelines, and related texts such as codes of practice under the Joint FAO/WHO Food Standards Programme.

Communication: Interchange of thoughts, opinions, or information, by means understood by both the sender and receiver.

Consumers: People who use products.

DNA: Deoxyribonucleic acid (sometimes called nucleic acid). A biological polymer that contains and transmits, through replication and transcription, the genetic information of the organism. It is composed of nucleotide units called bases (A-adenine, T-thymine, Q-guanosine, C-Cytosine). It is the specific order of these bases that can code instructions what the organism will be like.

Domestication (of plants and animals): Adaptation of wild plants and animals into forms that are useful to humans.

Ethics: Is the use of rational approach to examine and analyze moral concepts, questions, and resulting choices and actions in a specific area or situation. Whereas, what is considered as moral behavior may sometimes differ region to region, rules of ethical behavior should be universal. Thus, what may be moral may not entirely be ethical; however, what is ethical always contains a subset of moral concepts. In effect, ethics helps to define and incorporate the universal core of moral behavior. For a medical doctor, ethical rules mean, for example, to be helpful and do no harm, to respect a patient as a person, and to be nondiscriminatory.

Electroporation: Treatment of cells with an electrical current, resulting in the creation of temporary pores that allow an uptake of DNA into the cells.

Enzymes (restriction): A protein capable of catalyzing a reaction of a substrate.

Embryogenesis: Process of embryo formation.

Facts: Observations that can be measured, tested, and verified.

Functional foods: Is any food with health-promoting claims.

Gel electrophoresis: Is a technique whereby molecules are separated on the basis of their molecular weight and electric charge in a physical mixture (gel) submerged in a liquid medium. This technique is used, for example, to separate DNA, RNA, and proteins.

GM foods: Food that has been modified with the use of genetic engineering.

Genetic modification: Also termed "recombinant DNA technology" or "genetic engineering." These are technique that involves the isolation of genetic material, splice, alter, recombine and transfer it from one organism to another. The genetic material (DNA) has been altered in a way that does not occur naturally by mating or natural recombination. The use allows selected individual genes to be transferred from one organism into another, even between nonrelated species. These techniques can be performed at various levels: from whole genome manipulations through chromosome manipulations to precise modification of single genes. Genetic modification has come to include the manipulation and alteration of the genetic material of an organism in such a way as to allow it to produce proteins with properties different from those normally produced, or to produce entirely foreign proteins altogether.

Gene: Is the unit of heredity. It is encoded in the form of a DNA sequence.

Genetically modified organisms (GMOs): Organisms, which contain genetic information (usually one or more genes), that enriches its genome in a way which does not occur in nature.

Genotype: Genetic characteristics of an organism.

Genetic engineering: The formation of new combinations of hereditary material by processes that do not occur in nature. The technology is sometimes called modern biotechnology, gene technology, gene cloning or recombinant DNA technology and refers often to genetically modifying living organisms.

Genetic information: The sum total of hereditary that is needed for a species to survive generation to generation.

Genome: A complete set of genetic instructions of an organism. The instructions exist as specific sequences of DNA or RNA.

Heredity: Is the passing of specific characteristics (traits) from parents to offsprings.

Harm: A negative event that results in damage.

Hazard: Source of danger identified on the basis of some intrinsic properties or probability of occurrence.

Host: An organism harboring and supporting growth of another organism. The relationship could be parasitic (benefiting only of the two organisms) or symbiotic (benefiting both organisms).

Morality: A decision making process based on attitudes that help to distinguish correct and incorrect choices and actions, thus in the process defining the character of the individual, group, or a society. Morality is influenced by religion, regional societal values, beliefs, and "gut" feelings.

Microinjection: A process by which substances can be injected into cells using very small needles.

Nucleic acid: Composed of polynucleotides in which the nucleotide residues are linked in a specific sequence by phosphodiester bonds. It is usually a component of the DNA molecule.

Particle gun: A method by which DNA can be introduced into cells, based on the principle of shooting particles into cells coated with the DNA.

PCR (Polymerase Chain Reaction): A technique for amplifying (multiplying) short segments of DNA by repeated cycles of DNA synthesis.

Phenotype: Observable characteristic of an organism resulting from the expression of the organism's genes (its genotype).

Plant tissue culture: A technique to grow and differentiate plant cells in vitro, usually with the aim to regenerate a complete plant from single cells or clusters of cells.

Plasmid: Is a carrier DNA molecule that can replicate independently of the host chromosomal DNA. It is often used in genetic engineering to introduce foreign DNA into cells and help their replication therein. In such a case it is called a vector.

Proteins: A large molecule composed of amino acids. There are 20 amino acids that can form proteins. Any combination of amino acids can be used for the creation of proteins. This depends on a complex process that starts with the decision which genetic information of an organism is to be transcribed. Proteins are essential for the existence of living organisms. For example, all enzymes that enable cellular processes to proceed, are proteins.

Precaution: Prudent foresight; actions taken to ensure good results.

Probability: Likelihood of an event taking place.

Public: Concerning the people as a whole.

Risk: Exposure to danger, sometimes also defined as the probability of harm. The risk can be voluntary (accepting and knowing the dangers involved), nonvoluntary (not knowing the dangers) and involuntary (forced into a dangerous situation without consent).

Recombination: Exchange of genetic material (DNA or RNA) between two individual organisms, resulting in a changed genetic makeup and properties. The exchange is heritable and permanent.

Replication: Copying of the genetic material.

Risk: Describes the magnitude of harm caused by a hazard and the frequency with which that hazard occurs.

RNA: Ribonucleic acid (sometimes called nucleic acid). A biological polymer that is usually involved in transcribing DNA information that can lead to the formation of proteins. It is composed of the same nucleotide units as DNA except for thymine that is replaced by U-Uracil. In some organisms, such as viruses, RNA performs similar function as DNA – containing and transmitting the genetic information of the organism.

Ribosomes: Are complexes of RNA and proteins with a function to translate the genetic information of the cells into proteins.

Somatic (asexual) embryogenesis: Process of embryo formation without the involvement of sexual fertilization.

Substantial equivalence: Indication that the composition, nutritional value, or intended use of GM food has not been altered. If GMO products are substantially equivalent to the non-GM counterparts, they do not need to be labeled.

Synergism: The association of two or more viruses acting at one time and affecting a change which one only is not able to make.

Symptom: Visible or otherwise detectable phenotype abnormality arising from disease

Threat: An indication of something undesirable likely to happen.

Transcription: Transfer of genetic information, usually from DNA onto RNA.

Transgenic plants: Plants containing artificially transferred pieces of DNA from other living organisms by means of genetic engineering.

Transgene: A gene which has been transferred into another organism.

Uncertainty: Reduced confidence in estimating the likelihood of an event taking place (see also probability). Uncertainty can be of quantitative or qualitative nature.

Virus strain: A group of similar virus isolates, that are serologically or immunologically related.

Virus: (a latin word means poison) is an infectious submicroscopic and filterable noncellular agent that multiplies only in living cells and often causes diseases.

X-ray crystallography: Allows determination of the arrangement of atoms on the basis of their crystal structure.

References – Key Resources

Blaine K, Powell D (2001) Communication of food-related risks. *AgBio Forum* 4:179–185. The article can be downloaded from http://www.agbioforum.org

Coombs J (1986) Macmillan dictionary of Biotechnology. Macmillan books, London

Daniell H Streatfield SJ, Wycoff K (2001) Medical molecular farming: production of antibodies, biopharmaceuticals and edible vaccines in plants. *Trends Plant Sci* 6(5):219–226

Douma WT (2003) The precautionary principle: its application in international, European and Dutch law. T.M.C. Asser Press, The Hague

European Commission (2002) Scenarios for co-existence of genetically modified, conventional and organic crops in European agriculture. European Commission Report EUR 20394

Gaskell G, Bauer M, Durant J (eds) (2002) Biotechnology: the making of a global controversy ed. Cambridge University Press, Cambridge

Giddings G, Allison G, Brooks D, Carter A (2000) Transgenic plants as factories for biopharmaceuticals. *Nature Biotechnology* 18(9):1151–1155

Hodson A (1992) Essential Genetics. Bloomsbury, London

James C (2007) Global Review of Commercialised Transgenic Crops: ISAAA Briefs No36

Marris C (2001) Public views on GMOs: deconstructing the myths. *EMBO reports* 21:545–548

Pechan and deVries (2005) Genes On The Menu: Facts For Knowledge-Based Decisions, Springer Publishers, Heidelberg

Reiss M (2002) Labelling GM foods-the ethical way forward. *Nature Biotechnology* 20(9):868

Reiss M, Strangham S (1996) Improving nature? Cambridge University Press, Cambridge

Tait J (2001) More Faust than Frankenstein: The European debate about the precautionary principle and risk regulation of genetically modified crops. *Journal of Risk Research* 4:175–189

Thompson PB (2003) Value judgements and risk comparisons. The case of genetically engineered crops. *Plant Physiology* 132:10–16

Internet-Based References

Australian Office of Gene technology: World Agricultural Biotechnology on GMOs: What is Biotechnology? What Is Gene Technology? http://www.ogtr.gov.au/pdf/public/factwhatis.pdf

Codex Alimentarius: including minutes of all their meetings. http://www.codexalimentarius.net/agend.htm

Eurobarometer 51.1 "The Europeans and the Environment", conducted between April and May 1999, published in September 1999. http://europa.eu.int/comm/dg10/epo/eb.html

Eurobarometer 52.1 "The Europeans and Biotechnology", conducted between November and December 1999, published in March 2000. http://europa.eu.int/comm/dg10/epo/eb.html

European Commission: The website provides access to the web pages of all Directorate Generals and as well according to subjects such as food safety. http://europa.eu.int/comm/index_en.htm

Articles and links on Co-existence can be obtained for example under http://www.europa.eu.int/comm/food/risk

Communication on the Precautionary Principle COM (2000). http://www.Eur-lex/en/com/index.html

Positions paper for Codex Alimentarius CCGP-Codex Committee on General Principles 2000. http://www.Eur-lex/en/com/index.html

FAO (Food and Agriculture Organization). Contains wealth of agriculture related information. For example information on principles for the Risk Analysis of Foods Derived from Modern Biotechnogy. ftp://ftp.fao.org/codex/standard/en/CodexTextsBiotechFoods.pdf

German Federal Environmental Agency. Genetic Engineering and Organic Farming, Barth R et al. http://www.oeko.de/bereiche/gentech/documents/gruene_gentech_en.pdf

IPR helpdesk: source and guide to patent information in the European Union. http://www.cordis.lu/ipr-helpdesk/en/home.html

OECD consensus documents on biosafety issues, including a series of document related GM plants. http://www.oecd.org

Pew Initiative: Future Uses Of Agricultural Biotechnology, Michael Rodemeyer. http://pewagbiotech.org/research/harvest/harvest.pdf

TWN (third world network): is involved in issues relating to development, the Third World and North-South issues. Biotechnology and biosafety is one of important issues. http://www.twnside.org.sg/bio.htm

UK Patent Office home page. http://www.patent.gov.uk/

United Nations Environment Programme International Register on Biosafety: This Web site offers information from many sources on biosafety. http://www.chem.unep.ch/biodiv/

USDA (US Department of Agriculture): good source of information on GM crop usage. http://www.usda.gov/usda.htm. For example http://www.ers.usda.gov/Briefing/biotechnology/chapter1.htm. Includes also information on Plant Variety Protection Act in the US. http://www.ams.usda.gov/science/PVPO/PVPindex.htm

USEP (US Environmental Protection Agency): Bt plant pesticides risk and benefit assessments by the U.S. Environmental Protection Agency with links to background documents. http://www.foodsafetynetwork.ca and http://www.biotech-info.net/bt-transgenics.html

WTO: the global harmonisation of trade and intellectual property provision is sought through the World Trade Organisation, including issues such as TRIPS – the Agreement on Trade Related Intellectual Property Rights. http://www.wto.org

Index

P. Pechan et al., *Safe or Not Safe: Deciding What Risks to Accept in Our Environment and Food*, DOI 10.1007/978-1-4419-7868-4, © Springer Science+Business Media, LLC 2011